Positioning of Robotic Manipulator End-Effector Using Joint Error Maximum Mutual Compensation

Yauheni Veryha

2006

Bibliografische Information Der Deutschen Bibliothek

Die Deutsche Bibliothek verzeichnet diese Publikation in der Deutschen Nationalbibliografie; detaillierte bibliografische Daten sind im Internet über http://dnb.ddb.de abrufbar.

ISBN 3-8325-1343-4

Logos Verlag Berlin
Comeniushof, Gubener Str. 47,
10243 Berlin
Tel.: +49 030 42 85 10 90
Fax: +49 030 42 85 10 92
INTERNET: http://www.logos-verlag.de

ACKNOWLEDGMENTS

I would like to thank my family and, especially, my wife Kate, for their assistance and emotional support. Without them, I would never have come so far with this book. Special thanks go to Professor Henrik Gordon Petersen from the University of Southern Denmark and Professor Jerzy Kurek from the Warsaw University of Technology who continuously supported me with advices during the preparation of topics covered in this book.

CONTENTS

ABBREVIATIONS

2-R	- Two rotary joints
3-D	- Three-dimensional
3-R	- Three rotary joints
ABB	- Second largest engineering company in Europe after Siemens
AC	- Alternative Current
API	- Application Program Interface
BFGS	- Broyden, Fletcher, Goldfarb and Shanno
CAD	- Computer Aided Data
DC	- Direct Current
DOF	- Degrees of Freedom
GUI	- Graphical User Interface
ISO	- International Standardization Organization
MS	- Microsoft (leading international software development company)
OLP	- Off-Line Programming
PC	- Personal Computer
PUMA	- Programmable Universal Manipulator
RM 10-01	- SCARA type robot produced by Bulgarian robot manufacturing plant
RMS	- Root Mean Square
SCARA	- Selective Compliant Assembly Robot Arm
SYNE	- SYstematic Non-redundant and Extendable
VBA	- Visual Basic for Applications

PREFACE

Robots are widely used in our life, for example, in automotive industry, space flights, entertainment, nuclear industry, *etc.* Nowadays, you may hardly find any car that was not built with help of robots (see Fig. 0.1).

Fig. 0.1. Robots performing welding operations on car body

Such applications of robots as space exploration, handicapped aid and medical applications are becoming more and more common. Some interesting facts about robotics today are:

- *Market of industrial robots steadily grows (yearly ~ 3-10 %)*
    - International robotics manufacturers ship about USD $5 billion worth of products a year;
    - Automotive industry is one of the biggest consumers with a main focus on making robots cheaper, more reliable, accurate and flexible;
    - Robot applications in service industry, medical surgery and entertainment become more and more mature;
    - One cannot imagine space exploration without robots any more.
- *Forecasts for robots*
    - A Research Group estimated 2.1 million robots would be sold as common tools by 2006. (csmonitor.com, 02/05/04);
    - The United Nations forecasted that 4.1 million robots would be used worldwide by 2007 (Technology Review, 10/20/04);
    - Following the lead of Honda, Toyota plans to sell robots that serve the needs of families and the elderly by 2010. (NY Times, 05/31/05).

Many researchers have been doing their best in the last decades to improve such important characteristics of robots as their positioning and orientation accuracy. However, constantly increasing industry demand of achieving high positioning accuracy still requires further research and improvements in this area.

Have you ever tried to build the highest one-column tower out of domino blocks? If yes, then you know that even if on one of the levels of your tower you make a small mistake and place the domino block not perfectly in the middle, later on you may still have a chance to slightly compensate it by placing the next domino block slightly in the opposite direction of the tower decline caused by a wrong position of your previous domino block (see Fig. 0.2). Of course, it would be still better not to make that mistake at all and then your tower could become even higher than you have it now, but no one is perfect. Interestingly, if you have some imagination then you can see that a very similar effect can be observed in joints of industrial robots. If you imagine that the height of your domino tower is equivalent to the robot end-effector positioning accuracy, then the errors in robot joints can be equivalent to the inaccuracy of placing the domino blocks on the top of your tower. Hence, if the position of the following domino block in the tower can compensate previous mistakes made with other domino blocks, then one can expect that joint errors in robots can also compensate each other if one is able to find appropriate configurations for robot joints. The latter problem was targeted in this book. The general-purpose algorithms and simulation frameworks to find optimal configurations of various types of industrial robots are presented. The method for robotic manipulator positioning accuracy improvement using joint error maximum mutual compensation was developed to help engineers to design and implement their robotic systems with the maximum possible positioning and orientation accuracy. The practical approach presented in this book can be widely used on the stages of robot end-effector trajectory planning. In most practical cases, one can observe the improvement of robot end-effector positioning accuracy by ten - fifteen percent and, in extreme cases, by two times and more. Surprisingly, this can be achieved without any additional hardware or measurement equipment.

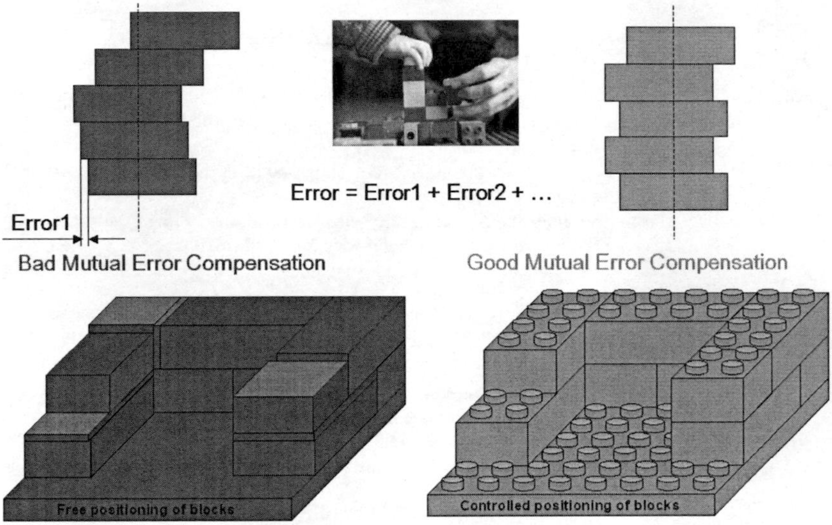

Fig. 0.2. Effect of mutual error compensation

To summarize the whole, we used such undesirable factors as robot joint errors to generate a positive effect of joint error maximum mutual compensation in industrial robots.

INTRODUCTION

> *"Any intelligent fool can make things bigger, more complex, and more violent. It takes a touch of genius - and a lot of courage - to move in the opposite direction."* **E.F. Schumacker**

Currently, a number of different types of robotic manipulators is used in the industrial setting for performing a variety of tasks, such as automated assembly, spray painting, sealing and other operations (see Fig. 0.3). Certain operations performed by industrial robots, such as laser cutting, high pressure water jet cutting, *etc.*, require fine accuracy in terms of robot end-effector path-tracking. Other operations, such as spot welding, some types of automated assembly and measuring require fine accuracy mostly in terms of robot end-effector positioning accuracy. In this book, we mostly deal with the latter applications of industrial robots that require fine accuracy in terms of robot end-effector positioning accuracy.

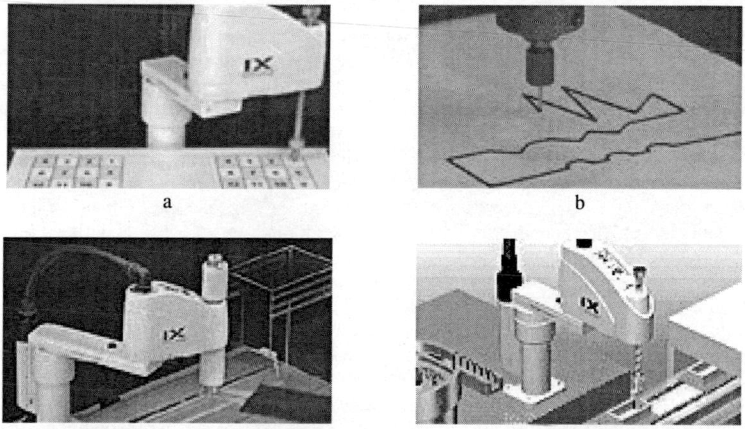

Fig. 0.3. SCARA robots executing the following technological operations:
a – palletizing; b – sealing; c –room cleaning; d – pick and place.

The accuracy of robotic manipulator end-effector positioning plays a key role for industrial robots. It defines the quality of performing given technological operations (*e.g.*, automated assembly, measurement, *etc.*). The attention to the problem of improving robotic manipulator positioning accuracy is motivated by industry demand to increase the accuracy of performed technological operations. End-effector positioning accuracy and positioning repeatability of industrial robots are one of the most important characteristics of robot end-effector positioning. These and other robot characteristics are defined in the industrial standard ISO 9283 (1998).

Robot end-effector positioning accuracy represents the result of vector sum of joint positioning errors appeared due to sensor inaccuracies, elastic vibrations, programming errors, coordinate transformation errors, *etc.* A number of performed laboratory experiments for robotic manipulators with rotary joints showed significant dependence of robot positioning accuracy in the robot workspace on robot end-effector location (Dimov, Dobrinov and

Boiadjiev, 1997; Kieffer, Cahill and James, 1997). The dependence of robot end-effector positioning accuracy on joint errors has a geometrical character. This allows locating special robot configurations and, thus, special areas in robot workspace with the highest robot end-effector positioning accuracy (Smolnikov, 1990). It is logical to propose that high-precision technological operations are executed by robot end-effector in the areas of the robot workspace with the highest robot end-effector positioning accuracy. This idea was used as the basis in this book to develop an effective method and framework for the implementation of joint error maximum mutual compensation in practice for conventional industrial robots. The research results are applicable for almost any type of industrial robots. The main robot prototypes, used in this research, were 2-R plane robotic manipulator, hexapod robot and PUMA - 560 type robotic manipulator.

High accuracy is most often achieved solely by designing the robot to be stiff and having little or no backlash. This means that joint errors are made small by design. However, to achieve say just a factor of two improvement in joint errors may increase the cost of the robot quite dramatically. Therefore, it is somewhat surprising how little effort there have been in searching for improvements by other methods apart from calibration.

The method of joint error maximum mutual compensation described in this book is applied after the calibration to compensate for the non-repetitive part of the joint errors. To be more specifically, consider a fixed robot configuration, where joint encoder readings are given by some values, and where the tool position and orientation has been measured by some external system. Ideally, this configuration can then be calibrated. However, when the robot is moved around and then back to the same encoder readings, the position and orientation of the tool will be different due to the non-repetitive part of the joint error. In the method of joint error maximum mutual compensation, one searches for optimal robot configurations where these unknown non-repetitive joint errors lead to as small an error as possible on the position and orientation of the end effector. This optimal configuration (or a neighbourhood around this configuration) can then be used for the required high precision tasks. Notice, that the method of joint error maximum mutual compensation can also be used without a prior calibration, in which case the joint errors must be expected to be bigger, but where one may then also have a better chance of making some vague estimates of the sign and size of these errors. It should also be noticed that there may be constraints on the search so that it is only admissible to use a limited part of the working area of the robot.

In the first chapter, the detailed overview of existing approaches to robot end-effector positioning improvement using error compensation is presented. In the second chapter, the method of robot end-effector positioning accuracy improvement using joint error maximum mutual compensation is presented. The general scheme, limitations, experiments and simulation results for 2-R SCARA type robotic manipulator are presented. In the third chapter, positioning accuracy improvement of redundant 3-R robotic manipulator end-effector and PUMA - 560 type robotic manipulator end-effector using robot joint error maximum mutual compensation is presented. In the forth chapter, the simulation of end-effector positioning accuracy improvement for hexapod robot using joint error maximum mutual compensation is presented. Typical robotic systems, where the developed framework of the joint error maximum compensation could be applied, are presented in the fifth chapter. Appendix A contains used MatLab program listings.

1. ROBOT END-EFFECTOR POSITIONING ACCURACY IMPROVEMENT

1.1. SURVEY OF ERROR COMPENSATION METHODS IN ROBOTIC MANIPULATORS

Some of the most important quality characteristics of an industrial robot are its repeatability and accuracy. Repeatability for robotic manipulators can be defined as the precision with which the manipulator is capable to return to a commanded point in the workspace. Accuracy for robotic manipulators is the measure of deviation in terms of position and orientation of the robot end-effector between the commanded path and the actually achieved path. Nowadays robots have a very high repeatability, but their accuracy is still relatively poor. The accuracy is up to ten times of the typical repeatability of industrial robots, which is about ± (0.1 – 1.0) mm.

An improvement of robot end-effector positioning accuracy is the subject of many research papers in the area of robotics. The theoretical basis of end-effector positioning accuracy research in robotic manipulators can be found in the works (Broderick and Cirpa, 1988; Slocum, 1992; Smolnikov, 1990). A geometrical dependence of end-effector positioning accuracy on joint error values and a possibility to use joint error mutual compensation for rotational joints were mentioned in those works. However, no input was provided on the practical aspects of the application of this geometrical dependence. The mathematical model of joint error mutual compensation was presented in (Smolnikov, 1990) for 2-R and 3-R robotic manipulators, but it was not extended for its application in practice. Dimov, Dobrinov and Boiadjiev (1997) presented experimental results of end-effector positioning accuracy and end-effector positioning repeatability measurements of SCARA type (RM 10-01) robotic manipulator with two rotational and one translational joints. The experiments were conducted in accordance with the industrial standard ISO 9283 (1998) and showed a significant dependence of end-effector positioning accuracy and positioning repeatability on the location of robot end-effector in the robot workspace.

Key methods of error compensation in robotic manipulators are shown in Fig. 1.1. Referring to Fig. 1.1, one should mention that conventional robot calibration methods usually experience problems:

- to compensate non-geometric effects such as gear backlash, tumbling, elastic deflections, *etc.* (Vincze et al., 1994);
- to compensate random errors;
- to compensate errors which are beyond the resolution of measuring equipment, encoders, *etc.*
- in some cases, because the calibration procedure is too complex and expensive (*e.g.*, time-consuming or requires very expensive measuring equipment).

Some of the cnconventional error compensation methods can help solving above-mentioned problems:

- Joint error mutual compensation (Smolnikov, 2000);
- Elimination of redundant error parameters (Meggiolaro *et al*, 1999);
- Joint error maximum mutual compensation (Veryha, 2003).

The accurate positioning of the robot end-effector can be obtained by appropriately modifying the mechanical structure of the robot, for example, by replacing worn-out components by new more accurately manufactured ones (Hayati and Mirmirani, 1985). However, in most cases, it is not enough. Kinematic robot calibration is one of the key requirements for the successful application of industrial robots (Bernhardt and Albright, 1993). To compensate for inaccurate robot end-effector positioning, offline generated poses need to be corrected using a calibrated kinematic model. This leads the robot to the desired poses. Positional calibration of robot end-effector can be generally categorized into hardware and software oriented methods (Bernhardt and Albright, 1993):

- Software methods
    - Re-mastering
    - Robot calibration (see Fig. 1.2)
- Mechanical adjustments.

The software-oriented methods of the positional calibration address positioning accuracy on robot controller (or robot application level). A generally supported software method falling into the category of software-oriented methods is known as re-mastering (Bernhardt and Albright, 1993). Re-mastering is a method where a joint is moved to a pre-defined position (usually the zero position designated, for example, by a mark on the neighbour link). The controller software is then updated with this new reference position by issuing the appropriate command to the robot controller.

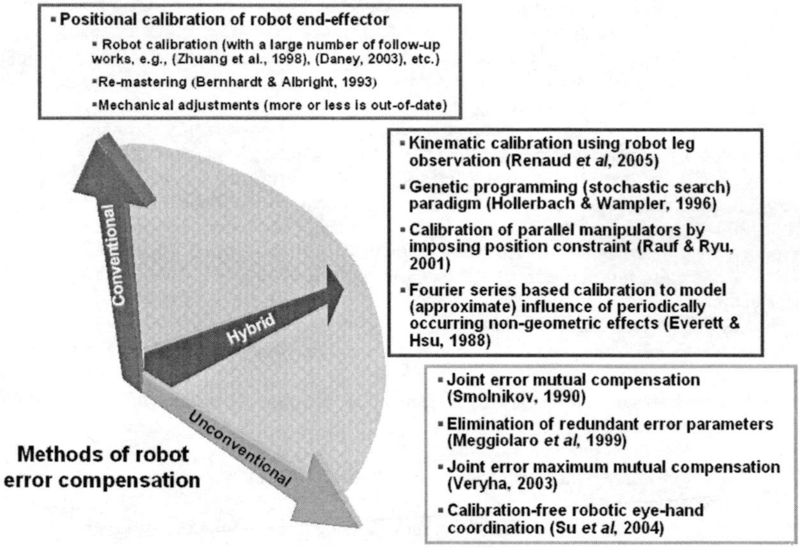

Fig. 1.1. Key methods of error compensation in robotic manipulators

Robot calibration (see Fig. 1.2) applies position compensation on robot application level based on a calibrated model. Robot calibration is a term associated with a set of software methods aiming at the identification of accurate robot models used with the objective to increase the positioning accuracy of the robot (Roth, Mooring and Ravani, 1987). Robot poses defined in programs, which may have been generated by OLP (Off-Line Programming) systems, are modified by subtracting the expected positioning error estimated by an accurate calibrated model. In this way, "false" target poses are produced with the objective to compensate for the positioning error. The deviation of the manipulator at these altered poses eventually leads the robot to the desired positions. In this method, the positions and the applied corrections are defined in the task space (Bernhardt and Albright, 1993).

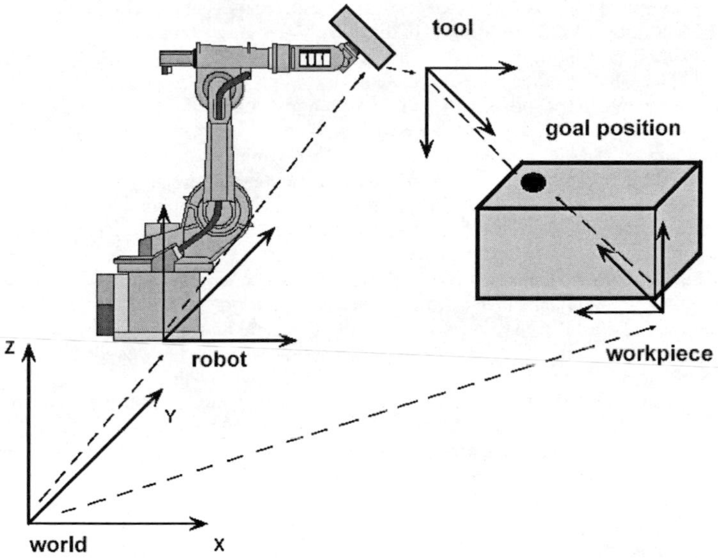

Fig. 1.2. Illustration of the robot calibration task in the simulation environment

According to the variety of tasks of industrial robots, there is a wide range of requirements for calibration methods (Diewald, Godding, and Henrich, 1993). Some of the most important requirements include:

- Accuracy potential of 0.1 mm or better;
- The cost/performance ratio should be optimal for the economical benefit of robot usage;
- Ease of use of the calibration method, which should not require special skills or too expensive equipment;
- The measurement system must be brought to the robot and not vice-versa;
- The measurement system should work under factory floor conditions and should not pose special requirements to the environment.

There is a distinction between static and dynamic calibration methods (Bernhardt and Albright, 1993). Static calibration aims at the identification of accurate models covering all physical properties and effects that influence the static positioning accuracy of the manipulator. Dynamic calibration builds upon the results of static calibration and addresses the identification of models describing motion characteristics of the manipulator (forces and actuator torques) and dynamic effects that occur on a manipulator such as friction, link stiffness, *etc.* To enable dynamic calibration, measurements of motion and forces of the manipulator are required. However, the difficulties in accurate tracking of these properties throughout the robot workspace and the complex problem of simultaneous identification of mass and friction parameters still limit the application of the dynamic calibration. Thus, most of the actual research works are concerned with the static kinematic robot calibration, which is typically carried out using the following steps:

1. Creation of a suitable kinematic model (usually based on prior engineering knowledge) that provides a model structure and nominal parameter values;
2. Measuring end-effector location of the manipulator in several positions;
3. Identification of the model parameters based on the measurements;
4. Implementation of the identified model.

One of the alternative approaches to the conventional kinematic robot calibration is the development of the inverse static kinematic calibration methods based on the genetic programming (stochastic search technique) paradigm (Hollerbach and Wampler, 1996). In such methods, the process of robot calibration is fully automated by applying symbolic model regression to model synthesis (structure and parameters) without involving iterative numerical methods for parameter identification, thus avoiding their drawbacks (*e.g.*, local convergence, numerical instability and parameter discontinuities). Genetic programming is employed in such methods to account, for example, for such effects like gear transmission errors, and to induce joint correction models used to compensate for positioning errors. Unlike with other numerical approaches that use, for example, artificial neural networks, genetic programming allows obtaining not only the required error compensation but also a symbolic description of the complex joint related effects, which enables further mathematical analysis.

Non-geometric effects are usually modeled by adding terms or more complex components to the overall geometric model of the manipulator. Since non-geometric effects are primarily due to joint related characteristics (Vincze *et al.*, 1994), the most common model adopted is a simple linear joint correction model (Hollerbach and Wampler, 1996) where the effective joint angle is computed from the joint angle sensor reading depending on the joint angle zero offset and joint transmission gain. This model is applied to each joint adding two more parameters per joint to be calibrated to the overall model. A comprehensive list of non-geometric errors and their models can be found in (Vincze *et al.*, 1999). Vincze *et al.* (1999) introduced a deterministic method of automatically generating robot models based on SYNE-axis (SYstematic Non-redundant and Extendable) description. This method composes the calibration model from a given geometric description of the manipulator and given non-geometric models to be used. To achieve non-redundancy of the composed model and to improve its accuracy, deterministic rules are applied to eliminate redundant parameters. If the accuracy after parameter identification is not sufficient, the modeling procedure may be repeated using a different set of non-geometric models.

When designing tasks for a particular robot using OLP systems, information from a calibrated kinematic model of the manipulator is used to correct designed (or nominal) poses with the expected positioning error to compensate for the effective positioning error. This correcting pose-to-pose mapping can be provided by a sequence of calibrated inverse and nominal forward kinematic model. Designed poses in task space are first transformed by the calibrated inverse model into calibrated joint configurations, which are then transformed back into task space. Alternatively, the calibrated joint angle poses could be fed directly into the robot controller saving two transformations. Typical measurement devices for robot calibration are wire potentiometers, telescopic ball system measured by a radial distance linear transducer, interferometer, ultrasonic systems (Berg, 1993), proximity sensors, imaging laser tracking systems (Vincze, Prenninger and Gander, 1994), single and stereo camera systems, magnetic trackers, *etc.* Different measurement systems are available varying in measurement method (contact and non-contact), the number of captured DOF's, accuracy and costs. To ensure a good performance of the parameter identification procedure, a sufficiently large set of data samples needs to be recorded where the poses have to be selected throughout the workspace in a way that guarantees the best observability of all parameters to be calibrated (Hollerbach and Wampler, 1996). To identify the influence of a parameter on all DOF's, it would in general be beneficial to involve full pose measurements. In practice, however, appropriate measurement devices may be quite expensive, relatively slow and difficult to set up. Alternatively, calibration can be performed using only position measurements, since all kinematic parameters of a manipulator may be identified based on position measurements if the measured points are not located along the end-effector axis.

Locating the manipulator home position is a common calibration technique, which can be divided into three main categories - relative, optimal and levelling based methods (Bernhardt and Albright, 1993). The home position of an industrial manipulator is a position where all joint angles have a pre-defined value (*e.g.*, 0 or 90 degrees), which can be transformed into Cartesian space via the robot kinematics. Large industrial manipulators require an accurately defined home position that can be restored with repeatability in the order of 0.1 mm in Cartesian space or 0.01 degrees in joint space. Relative calibration methods locate each robot link in a predefined position, relative to the previous link in the kinematic structure. This requires highly accurate robot parts and, thus, become an expensive option for high-volume manipulators. Optimal methods require a measurement system to measure a set of robot poses and use mathematical models of the robot kinematics to determine the link angles. The resulting home position is sensitive to the chosen calibration poses, leading to difficulties in maintaining a repeatable calibration. Leveling based methods utilize inclinometers to position each link parallel or perpendicular to the gravity vector and provide a simple, accurate and repeatable recalibration of the robot's home position. To achieve the required accuracy using a leveling system, several key accuracy and repeatability aspects (*e.g.*, the robot control system, the inclinometer unit and the mounting interface to the robot) must be considered. An example of the leveling system is a system that uses two sensors mounted on separate right-angle plates that are placed sequentially on various mounting plates bolted to the robot structure. Each joint of the robot is then manually jogged until the robot stands in a home position that is parallel and perpendicular to the gravity vector.

The last step in robot calibration typically involves all procedures and mechanisms necessary to transfer the calibration results into practice. In the offline programming, this includes the implementation of a postprocessor that uses information from the calibrated model to perform corrections on positional data in program files which have been generated by OLP systems (Hollerbach and Wampler, 1996). Basic numerical calibration methods and postprocessors have become built-in components in most of the recent OLP systems.

To specify the limitations of the calibration systems, one should mention that the robot control system is a function of the accuracy of the motor resolvers and can be expressed as the repeatability of the manipulator. The repeatability of the manipulator is typically of the order of 100 micrometers as defined by various ISO standards (*e.g.*, ISO 9283 (1998)), which test the robot at a defined speed and with a defined payload. This value cannot be compensated for, in most cases, using existing calibration system and can be considered to be a stochastic error around the chosen home position (Bernhardt and Albright, 1993).

As for the calibration of parallel robots, which become more and more popular in practice, one should mention self calibration. The purpose of self-calibration is to calibrate the closed-loop mechanism by using built-in sensors in the passive joints. After self-calibration, only the kinematic errors in the fixed transformations from the robot world frame to robot base frame and from the mobile platform frame to the end-effector frame need to be identified (Rauf and Ryu, 2001). Representative works in self-calibration are presented by Zhuang, Yan and Masory (1998) and Yiu, Meng and Li (2003). In (Zhuang, Yan and Masory, 1998), a self-calibration method is proposed for the conventional six-legged Stewart platform through the installation of redundant sensors in several passive joints and constructing a measurement residues with measured values and the computed values of these readable passive joint angles. When these passive joint angles are recorded at a sufficient number of measurement configurations, the actual kinematic parameters can be estimated by minimizing the measurement residues. Since this model is based on the linearization of the kinematic constraint equations, it converges rapidly (Zhuang, Yan and Masory, 1998). Iurascu and Park (1999) developed a unified geometric framework for the calibration of kinematic chains containing closed loops. Both joint encoder readings and end-effector pose measurements can be uniformly included into this framework. As a result, the kinematic calibration is cast as a nonlinear constrained optimization problem. Khalil and Besnard (1999) proposed an approach for self-calibration of parallel robots without extra sensors by measuring end-effector location of the manipulator in several positions and then identifying model parameters based on the measurements.

Robot calibration based on the implicit model is a standard method to calibrate parallel robots (Zhuang and Roth, 1993; Nahvi, Hollerbach and Hayward, 1994). For mathematical reasons, the number of equations given by the measurements in robot calibration has to be at least as large as the number of unknown parameters. Since the measurement data are usually given by sensors, it is necessary to take into account the noise associated with measuring devices. Some of the latest methods proposed by researchers attempted to handle the problem of partial information in the calibration of parallel robots by introducing semi-parametric parallel robot models (Daney and Emiris, 2001; Chai, Young and Tuersley, 2002), which lead to low computational effort important for real applications. Daney, Papegay and Neumaier (2004) proposed a method that gives a certified approximation of kinematic calibration. For a set of measurements given with attached uncertainties, a list of intervals for the kinematics parameters is returned, such that any solution corresponding to an instance of configuration satisfying the measurements has to belong to those intervals. This method is a new version based on interval arithmetic, using interval analysis of the so-called implicit or inverse calibration method, the most studied method for the identification of the kinematic parameters of a parallel robot (Zhuang, Yan and Masory, 1998; Daney, 2003).

An error function based on the inverse kinematic model is often considered as the most numerically efficient method (Besnard and Khalil, 2001), because an inverse kinematic model can be written in closed-form for most parallel structures, contrary to the direct kinematic model (Angeles, 2002). The main limitation is the necessary end-effector full-pose measurement (position and orientation of the end-effector), which have to be obtained using high-precision measuring equipment, which is usually quite expensive. The high-precision camera is one of the possible measuring devices. Not only end-effector observations, but also legs observation with a camera enable one to get attained poses and then achieve the mechanism kinematic calibration (Renaud *et al.*, 2004; Andreff, Marchadier and Martinet, 2005). For many parallel mechanisms, the simultaneous observation of the mechanism legs and end-effector can be achieved (Renaud *et al.*, 2004). As information redundancy is the basis for the error function design, the calibration efficiency should be logically improved by a simultaneous observation of all the mechanism elements (Andreff, Marchadier and Martinet, 2005).

As already mentioned, typically, robot end-effector positioning accuracy can be improved either by precise manufacturing and assembly or by calibration of each individual robot using a kinematic model which takes errors in robot structure into account. The latter has the advantage of leading to low cost solutions but requires sophisticated modeling of the robot structure, as described above. Another approach, which is not widely used in practice, is the effect of joint error maximum mutual compensation in robot structures (Veryha and Kurek, 2002; Veryha and Kurek, 2003), which will be covered in details in this book. Joint error maximum mutual compensation is very easy to use in practice, because it does not require any additional equipment and satisfies most of the above-mentioned requirements to robot calibration coming from practice. Joint error maximum mutual compensation can be used after the calibration of the given robotic manipulator is finished. Thus, it is not an alternative to the calibration; it is an add-on to calibration to further improve robot end-effector positioning accuracy even if joint errors are unknown or not precisely known. Joint error maximum mutual compensation can be used for both robotic manipulators with closed-loop control systems and for robotic manipulators with open-loop control systems.

1.2. PROBLEM FORMULATION

The main goal of this book is to present the results of our investigation (see Fig. 1.3) of opportunities in joint error maximum mutual compensation for improving robot end-effector positioning accuracy characteristics.

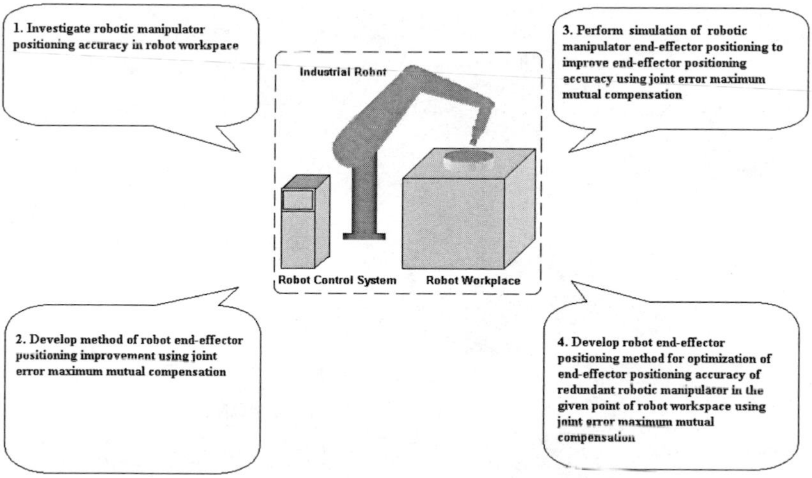

Fig. 1.3. Main research tasks for robot end-effector positioning improvement using joint error maximum mutual compensation

In our research, we used only robotic manipulators with rotational and linear joints. Two-dimensional (XY, XZ or YZ planes) joints can be usually presented as a composition of basic linear types. Thus, we did not use the latter type of joints in the given research. We did not use spherical joints in our research as well because their geometry, kinematics and dynamics differ largely from that of rotational joints and they are quite rarely used in conventional robotic manipulators. As a result, in this book we concentrated only on robotic manipulators with rotational and translational joints.

In practice, it is usually difficult to calculate or approximate joint errors of robotic manipulators with high precision because of the lack of appropriate high-precision sensors and/or open-loop control systems. A number of joint errors, like those caused by wearing of joint surfaces, mechanical defects and inaccuracies of joints, programming errors, *etc.* are hard to take into account in robot control system. In most conventional robotic manipulators, joint errors are only partly compensated, because one can not obtain precise values of joint errors using available sensors or approximation models of robotic manipulators. In this work, we propose to use joint error maximum mutual compensation to improve end-effector positioning accuracy of robotic manipulator with unknown or not precisely known joint error values. This method can be used for both robotic manipulators with closed-loop control systems and for robotic manipulators with open-loop control systems.

We think that the best effect of using joint error maximum mutual compensation can be achieved in robotic manipulators with open-loop control systems with stepping drives because they do not have feedback sensors and rely on the precision of the execution of control pulses coming from control system to the rotor or stator of stepping motor. This includes the application of special robots for manufacturing of complex high-precision mechanics, for example for optical devices, micro-electronics, *etc.* One of the special types of robotic manipulators that can be used for such operations is the hexapod. A hexapod is a highly rigid and dextrous parallel mechanism with 6 degrees of freedom which is a member of the Stewart-Gough family of platforms (Angeles, 2002). These mechanisms were originally designed for applications as diverse as tire testing and flight simulators. However, the basic properties of hexapod robot make it close to ideal for micro-handling purpose (Doering, 2004). By using a miniature version of the hexapod (see Fig. 1.4), we get the necessary dexterity to perform arbitrary motions in 3-D space and the rigidity to obtain the accuracy that we need.

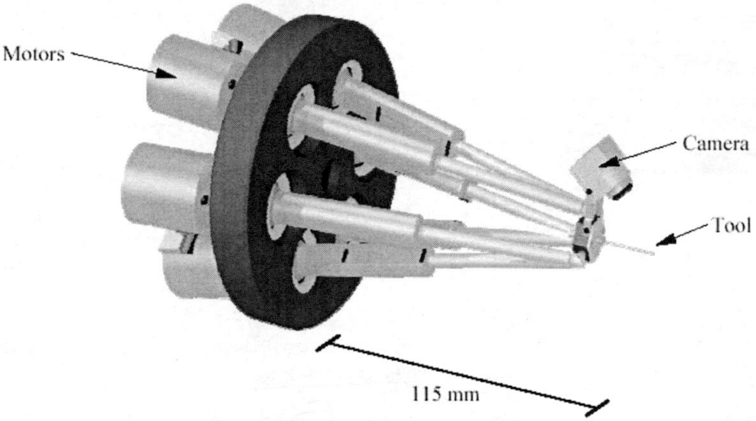

Fig. 1.4. Example of miniature hexapod

Obtaining the necessary accuracy is a difficult task requiring not only development of highly specialized and accurate mechanics but also the advanced control. The control of the hexapod relies on feedback provided by an unusual combination of sensor systems (Doering, 2004). In short, two different types of sensors are used. An external optical sensor system that measures the pose of the tool of the hexapod relative to some global workcell reference. In addition, a little video camera is mounted near the tool of the hexapod, as shown in Fig. 1.4. When the hexapod approaches the expected position of a certain workpiece, the object enters the field of view of the camera. Using advanced computer vision algorithms the video stream is analyzed and the pose of the tool relative to relevant features of the workpiece is calculated (Doering, 2004). These two measurements work in conjunction to provide highly accurate probing of the pose of the hexapod tool in order to enable relevant control.

We propose to use joint error maximum mutual compensation in hexapods to find such configurations of hexapods in which the best end-effector positioning accuracy can be achieved. The hexapod kinematics and application of joint error maximum mutual compensation for hexapods will be discussed in details in Chapter 4.

1.3. CLASSIFICATION AND IDENTIFICATION OF ERRORS IN ROBOTIC MANIPULATORS

In this sub-chapter, we will shortly present a classification of errors in robotic manipulators to identify the types of errors that we will mutually compensate in further chapters. There are many possible sources of errors in robotic manipulators. The main errors in robotic manipulators can be classified as (Slocum, 1992):

- *Mechanical system errors* (These errors result from machining and assembly tolerances of various mechanical components in robotic manipulators);
- *Deflections* (elastic deformations of mechanical components of robotic manipulators under load);
- *Measurement and control errors* (for example, the resolution of encoders and stepper motors);
- *Errors in joint components* (These errors include bearing run-out in rotating joints, rail curvature in linear joints, backlash in robotic manipulator joints and actuator gear box).

The scheme in Fig. 1.5 presents an overview of main errors in robotic manipulators. In most cases, the above-mentioned errors are relatively small but their effect on the end-effector of the robotic manipulator can be large (Slocum, 1992).

Further, errors can be distinguished into repeatable and random ones (Slocum, 1992). Repeatable errors are those which numerical value and sign are constant for a given robotic manipulator configuration. An example of a repeatable error is an assembly error. Random errors are those which numerical value or sign changes unpredictably. At each robotic manipulator configuration the exact magnitude and direction of random errors cannot be uniquely determined. It can be only specified over a range of values. An example of a random error is the error that occurs due to a backlash of an actuator gear train. Unfortunately, random values cannot be compensated using classical calibration techniques. Classical kinematic calibration and correction can only deal with repeatable errors (Slocum, 1992).

To describe the kinematics of a robotic manipulator, one has to define reference frames at the robotic manipulator base, end-effector and each of the joints characterized by the Denavit and Hartenberg parameters (Shahinpur, 1990). The position and orientation of a reference frame F_i with respect to the previous reference frame F_{i-1} is defined with a 4 x 4 matrix $\mathbf{A}_i$ that has a general form:

$$\mathbf{A}_i = \begin{bmatrix} \mathbf{R}_i & \mathbf{T}_i \\ 0 & 1 \end{bmatrix}, \qquad (1.1)$$

where $\mathbf{R}_i$ is a 3 x 3 orientation matrix composed of the direction cosines of frame F_i with respect to frame F_{i-1} and $\mathbf{T}_i$ is a 3x1 vector of coordinates of center O_i of frame F_i in F_{i-1} (see

Fig. 1.6). Elements of matrices $\mathbf{A}_i$ depend on the geometric parameters of robotic manipulators and joint coordinates $\mathbf{q}$.

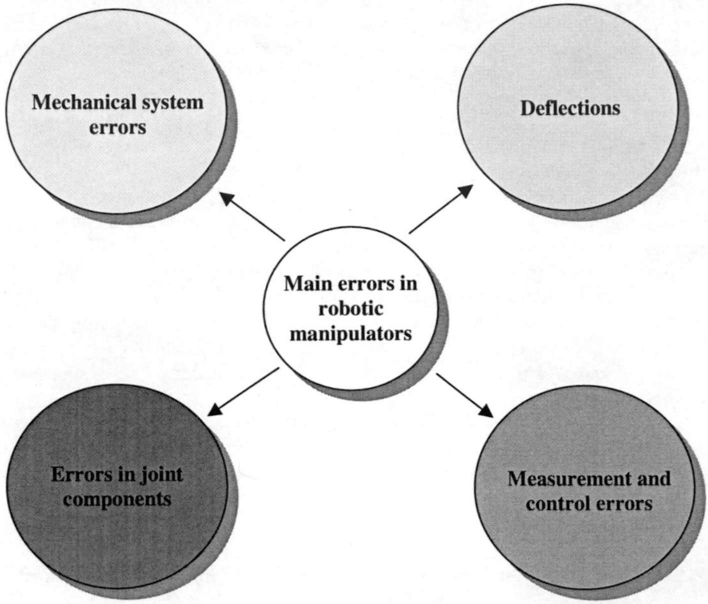

Fig. 1.5. Classification of errors in robotic manipulators

Errors change the geometric properties of a robotic manipulator. As a result, the frames defined at the manipulator joints are slightly displaced from their pre-defined ideal locations. Since the generalized errors e_{si}, e_{ri}, e_{pi}, e_{xi}, e_{yi} and e_{zi} are quite small, a first order approximation can be applied to their trigonometric functions and products obtaining (1.2) as a result (Shahinpur, 1990). The real position and orientation of a frame F_i^r (superscript r means *real*) with respect to its ideal location F_i^i (superscript i means *ideal*) can be represented by a 4 x 4 homogenous matrix $\mathbf{E}_i$ (Shahinpur, 1990):

$$\mathbf{E}_i = \begin{bmatrix} 1 & -e_{ri} & e_{si} & e_{xi} \\ e_{ri} & 1 & -e_{pi} & e_{yi} \\ -e_{si} & e_{pi} & 1 & e_{zi} \\ 0 & 0 & 0 & 1 \end{bmatrix}. \qquad (1.2)$$

The rotation part of matrix $\mathbf{E}_i$ (see (1.2)) is the result of the product of three consecutive rotations e_{si}, e_{ri} and e_{pi} around the Y, Z and X axes respectively. The subscripts s, r and p represent spin (yaw), roll and pitch, respectively (Euler angles (Shahinpur, 1990)). The translational part of matrix $\mathbf{E}_i$ is composed of the three coordinates e_{xi}, e_{yi} and e_{zi} of point O_i^r

in F_i^i. The generalized errors e_{si}, e_{ri}, e_{pi}, e_{xi}, e_{yi} and e_{zi} represent all robotic manipulator errors and can be calculated link by link.

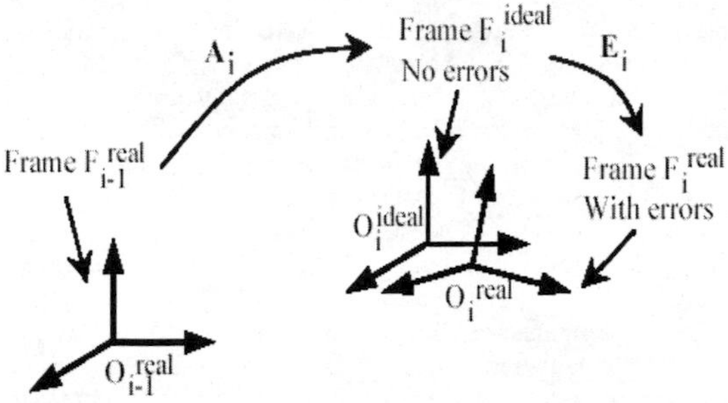

Fig. 1.6. Frames defined at robotic manipulator joints

The end-effector position and orientation error $\Delta\mathbf{X}$ is defined as 6 x 1 vector that represents the difference between the real position and orientation of the end-effector and the ideal (desired) one:

$$\Delta\mathbf{X} = \mathbf{X}^r - \mathbf{X}^i , \tag{1.3}$$

where $\mathbf{X}^r$ and $\mathbf{X}^i$ are the 6 x 1 vectors composed of the three positions and three orientations of the end-effector reference frame F_n in the initial reference system F_0, respectively, for the real and ideal cases (Shahinpur, 1990). When the generalized errors are considered in the model, the manipulator loop closure equation takes the form:

$$\mathbf{A}(\mathbf{q}, \boldsymbol{\varepsilon}, \mathbf{s}) = \mathbf{A}_1 \mathbf{E}_1 \mathbf{A}_2 \mathbf{E}_2 \ldots \mathbf{A}_n \mathbf{E}_n , \tag{1.4}$$

where $\mathbf{A}$ is a 4 x 4 homogeneous matrix of the type shown in equation (1.1) that describes the position and orientation of the end-effector frame F_n with respect to the inertial reference frame F_0 as a function of joint coordinates $\mathbf{q}$, generalized errors $\boldsymbol{\varepsilon}$ and structural parameters $\mathbf{s}$. The vector $\mathbf{X}^r$ can be written in a general form (Mirman and Gupta, 1993):

$$\mathbf{X}^r = \mathbf{f}^r (\mathbf{q}, \boldsymbol{\varepsilon}, \mathbf{s}) , \tag{1.5}$$

where $\mathbf{f}^r$ is a vector non-linear function of $\mathbf{q}$, $\boldsymbol{\varepsilon}$ and $\mathbf{s}$.

Since the generalized errors are relatively small, $\Delta\mathbf{X}$ can be calculated by the following linear equation (Shahinpur, 1990):

$$\Delta\mathbf{X} = \mathbf{J}_e \, \boldsymbol{\varepsilon} , \tag{1.6}$$

where $\mathbf{J}_e$ is the 6 x 6n Jacobian matrix (n is the number of joints) of the function $\mathbf{f}^r$ with respect to the elements of the generalized error vector $\boldsymbol{\varepsilon}$. The elements of $\mathbf{J}_e$ are defined as (Shahinpur, 1990):

$$\mathbf{J}_e[i, \quad j] = \frac{\partial \mathbf{f}^r[i]}{\partial \boldsymbol{\varepsilon}[j]} , \tag{1.7}$$

The value of i ranges from 1 to 6 and j ranges from 1 to $6n$. In general, $\mathbf{J}_e$ depends on the system configuration, geometry and weight (if there are elastic deflections in the system).

If the generalized errors $\boldsymbol{\varepsilon}$ are known then the end-effector position and orientation errors can be calculated using (1.6). Fig. 1.7 shows how an error model of the type of equation (1.6) can be used in a typical error compensation algorithm (Hollerbach and Wampler, 1996). The first step in the identification of generalized errors $\boldsymbol{\varepsilon}$ is the use of off-line measurements. The identification process to calculate $\boldsymbol{\varepsilon}$ (see Fig. 1.7) is based on the assumption that some components of vector $\Delta\mathbf{X}$ can be obtained experimentally at a finite number of different manipulator configurations (Mirman and Gupta, 1993). However, since position coordinates are much easier to measure in practice than orientations, in many cases only three position coordinates of $\Delta\mathbf{X}$ are measured, requiring twice the number of measurements for the calculation (Mirman and Gupta, 1993). Assuming that all six components of $\Delta\mathbf{X}$ can be measured, for an nth degree of freedom robotic manipulator, its 6n generalized errors $\boldsymbol{\varepsilon}$ can be calculated by fully measuring vector $\Delta\mathbf{X}$ at n different configurations and then writing equation (1.6) n times:

$$\Delta\mathbf{X}_t = \begin{bmatrix} \Delta\mathbf{X}_1 \\ \Delta\mathbf{X}_2 \\ \dots \\ \Delta\mathbf{X}_n \end{bmatrix} = \begin{bmatrix} \mathbf{J}_e(q_1, w_1) \\ \mathbf{J}_e(q_2, w_2) \\ \dots \\ \mathbf{J}_e(q_n, w_n) \end{bmatrix} \boldsymbol{\varepsilon} = \mathbf{J}_t \boldsymbol{\varepsilon} \quad , \tag{1.8}$$

where $\Delta\mathbf{X}_t$ is the $6n$ x 1 vector formed by all measured vectors $\Delta\mathbf{X}$ at the n different configurations and $\mathbf{J}_t$ is the $6n$ x $6n$ total Jacobian matrix formed by the n error Jacobian matrices at the n configurations (Borm and Menq, 1991).

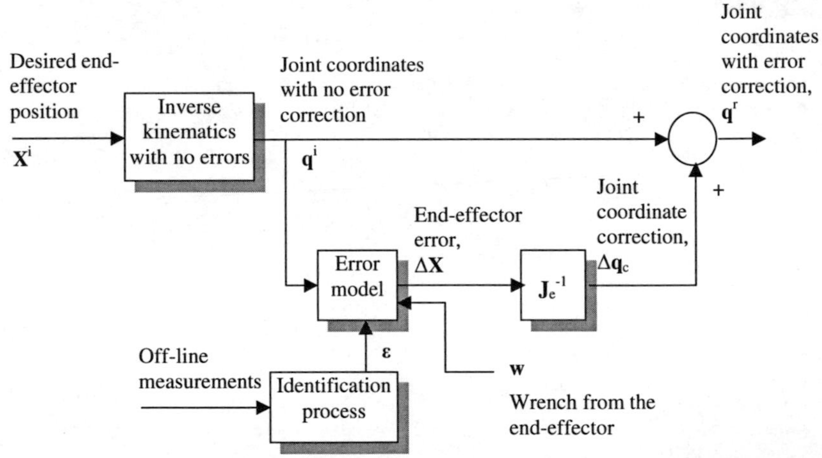

Fig. 1.7. Typical error compensation scheme

If matrix $\mathbf{J}_t$ is non-singular and the generalized errors $\boldsymbol{\varepsilon}$ do not depend on the configuration (which is normally not the case in practice), then $\boldsymbol{\varepsilon}$ is obtained simply by inverting $\mathbf{J}_t$:

$$\boldsymbol{\varepsilon} = \mathbf{J}_t^{-1}\,\Delta\mathbf{X}_t\,. \tag{1.9}$$

If $\mathbf{J}_t$ is singular then it is clear that equation (1.9) cannot be applied (Shahinpur, 1990). This can occur if some of the generalized errors ε_i result in end-effector errors in the same direction. By measuring this end-effector error it is not possible to distinguish the amount of the error contributed by each generalized ε_i (Vaichav and Magrab, 1987). This condition usually occurs because of the existence of special geometric conditions between the manipulator joint axes (as parallel or orthogonal axes) or the existence of prismatic joints (Vaichav and Magrab, 1987).

Partial measurement of vector $\Delta\mathbf{X}$, such as measuring only the position but not the orientation of the end-effector, can also lead to a singular $\mathbf{J}_t$. In this case, only linear combinations of generalized errors ε_i can be calculated. Mathematically, the singularity of $\mathbf{J}_t$ is expressed with a linear dependency of columns of $\mathbf{J}_t$. To find $\boldsymbol{\varepsilon}$ for a singular $\mathbf{J}_t$, one can use the procedure of reduction of $\mathbf{J}_t$ to a non-singular matrix (Borm and Menq, 1991; Vaichav and Magrab, 1987). To find $\boldsymbol{\varepsilon}$ when $\boldsymbol{\varepsilon} = \boldsymbol{\varepsilon}(\mathbf{q})$ one can use the polynomial approximation of generalized errors (Mirman and Gupta, 1993).

Unfortunately, the above-presented joint error compensation scheme is applicable only if joint errors are known or predictable with a high level of accuracy. In many industrial applications it is not the case. Thus, some other approaches to improve end-effector positioning accuracy are needed in practice for industrial robots with unknown or partly known joint errors. One of them is the method of joint error maximum mutual compensation that will be presented in next chapters.

2. ROBOT END-EFFECTOR POSITIONING ACCURACY IMPROVEMENT USING JOINT ERROR MAXIMUM MUTUAL COMPENSATION

2.1. ROBOT POSITIONING ACCURACY

One of the ways to improve robot end-effector positioning accuracy is to search for the optimal robot joint configurations (that provide the best end-effector positioning accuracy) in the robot workspace for the given end-effector trajectory (Smolnikov, 1990). As soon as such robot configurations are identified, high precision technological operations can be performed in those areas of robot workspace. The existence of optimal joint configurations of robotic manipulators with rotational joints is connected with the kinematic chains of the robotic manipulators (Smolnikov, 1990). There is a dependence of the robot end-effector positioning accuracy in the robot workspace on the robot end-effector location. The resulting robot end-effector positioning accuracy is defined by the vector sum of joint errors (Vaichav and Magrab, 1987; Smolnikov, 1990). The optimal robot configuration with the joint error maximum mutual compensation could improve the resulting robot end-effector positioning accuracy significantly.

Generally, there are two possibilities to search for the optimal robot configuration: global search in the whole robot workspace and local search in the given small area of the robot workspace. In both cases, the search should be done in advance, before accommodating other technological equipment and fixing assembly parts in their positions. This early investigation of the robot end-effector positioning accuracy and robot joint errors provides more flexibility in the later planning of robot operations with the optimal use of end-effector positioning accuracy.

To compensate joint errors of robots with rotational joints, there are basically two approaches, the same as with the control of robots (Broderick and Cirpa, 1988; Wehn and Belanger, 1997). One can either compute the Cartesian position and orientation by using the readily available high precision joint displacement measurements as arguments of the kinematic function or one can try to measure the taskspace position directly with additional sensors. Generally, both mentioned approaches could be used in order to provide the joint error maximum mutual compensation. However, the most effective could be the latter one. This method allows using the true kinematic function of the robotic manipulator and, therefore, providing significant end-effector positioning accuracy improvement. The drawback of the direct end-effector position measurement in 3-D space is expensive equipment required for 3-D high-precision measurements.

Robot end-effector positioning accuracy is defined by the industrial standard ISO 9283 (1998). Basic terms and their definitions from ISO 9283 (1998) are presented below to define robot end-effector positioning accuracy. First, before defining robot end-effector positioning accuracy, one has to consider the term "pose accuracy". Pose accuracy expresses the deviation between a command pose and the mean of the attained poses when approaching the command pose from the same direction (ISO 9283, 1998). Pose accuracy and repeatability characteristics quantify the differences that occur between a command and attained pose and the fluctuations in the attained poses for a series of repeat visits to a command pose (ISO 9283, 1998). The pose errors may be caused by internal control definitions, coordinate transformation errors, differences between the dimensions of the articulated structure and those used in the robot control system model, mechanical faults such as clearances, hysteresis, friction, and external influences such as temperature (ISO 9283, 1998).

The terms "command pose" and "attained pose", which are used to define pose accuracy, are explained using scheme shown in Fig. 2.1. Command pose is a pose specified through teach programming, numerical data entry through manual data input or off-line programming (see Fig. 2.1, x_c, y_c and z_c are Cartesian coordinates of the command pose) (ISO 9283, 1998). The command pose for teach programmed robots is to be defined as the measurement point (see Fig. 2.1) (ISO 9283, 1998). This point is reached during programming by moving the robot as close as possible to the defined points in the cube (see Fig. 2.1) (ISO 9283, 1998). The coordinates registered on the measuring system are then used as "command pose" when calculating accuracy based on the consecutive attained poses (ISO 9283, 1998).

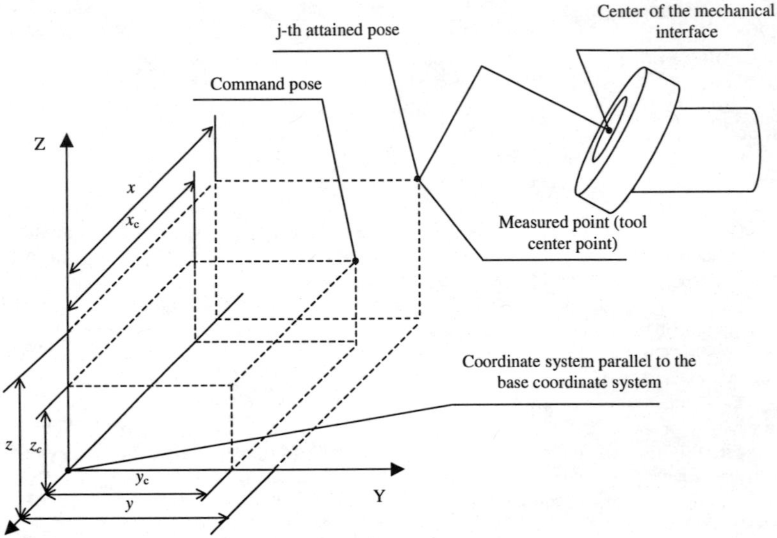

Fig. 2.1. Relation between command and attained pose (ISO 9283, 1998)

Attained pose (see Fig. 2.1, x, y and z are Cartesian coordinates of the attained pose) is a pose achieved by the robot under automatic mode in response to the command pose (ISO 9283, 1998).

Pose accuracy consists of the positioning accuracy and orientation accuracy (ISO 9283, 1998). Positioning accuracy is the difference between the position of a command pose and the barycenter of the attained positions (see Fig. 2.2) (ISO 9283, 1998). Another important characteristic of end-effector accuracy is "pose repeatability" which expresses the closeness of agreement between the attained poses after n repeat visits to the same command pose in the same direction (see Fig. 2.2). Pose repeatability is expressed by the positioning repeatability and orientation repeatability (ISO 9283, 1998).

To investigate possible values of positioning accuracy in real robotic manipulators, we will use the experimental data presented in (Dimov, Dobrinov and Boiadjiev, 1997) and make our own advanced analysis of these experimental results.

Dimov, Dobrinov and Boiadjiev (1997) presented their experimental results of measurements of positioning accuracy and positioning repeatability for the industrial robot

RM 10-01 (SCARA type robot produced by Bulgarian robot manufacturing plant) with two rotational joints in the horizontal plane and one translational joint in the vertical plane. The measurements were performed in accordance with the industrial standard ISO 9283 (1998) in five points of the robot workspace using electronic apparatus for linear measurements with analogue output - "TESA MODULE" (Dimov, Dobrinov and Boiadjiev, 1997). Points $\mathbf{P}_2$, $\mathbf{P}_3$, $\mathbf{P}_4$ and $\mathbf{P}_5$ were located in the corners of the abstract square (length 0.4 m) in the robot working plane. Point $\mathbf{P}_1$ was located in the diagonals' crossing of the abstract square. Coordinates of the points are presented in Table 2.1 (Dimov, Dobrinov and Boiadjiev, 1997). The scheme of the workspace of the robot RM 10-01 (link lengths $l_1 = 0.4$ m and $l_2 = 0.25$ m), where the measurements have been performed, is shown in Fig. 2.3 (Dimov, Dobrinov and Boiadjiev, 1997).

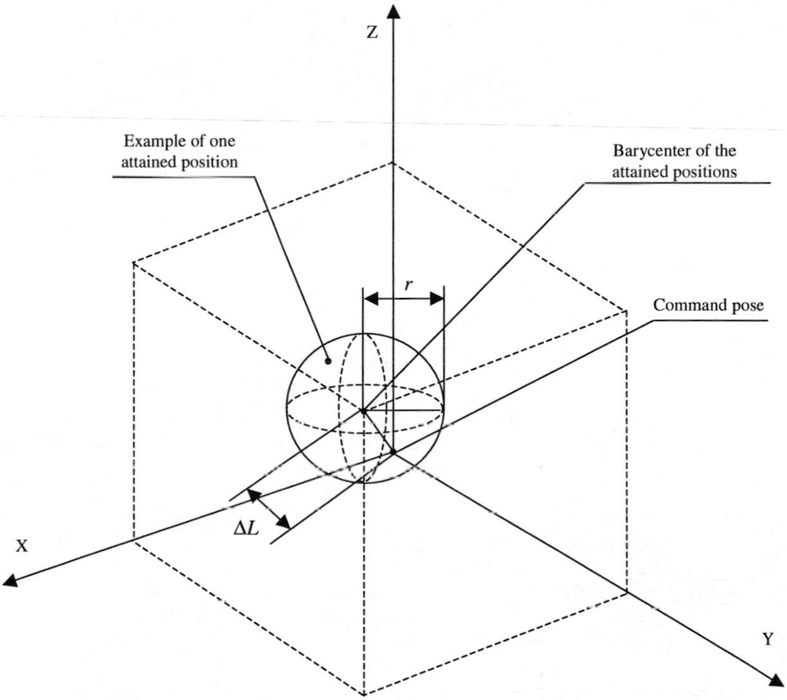

Fig. 2.2. Positioning accuracy and repeatability (ISO 9283, 1998)

The following symbols were used in Fig. 2.3: $\mathbf{P}_1$, $\mathbf{P}_2$, $\mathbf{P}_3$, $\mathbf{P}_4$ and $\mathbf{P}_5$ – points of end-effector positioning during an execution of the given trajectory; 1 and 2 – accordingly, first and second links of the robotic manipulator (the linear joint is not presented in the given scheme); x_0y_0 – base robot coordinate system; q_1 and q_2 – first and second robot joint coordinates. The robot end-effector consequently moved from point $\mathbf{P}_1$ to point $\mathbf{P}_2$, further in points $\mathbf{P}_3$, $\mathbf{P}_4$, $\mathbf{P}_5$ and returned back into the initial point $\mathbf{P}_1$ (Dimov, Dobrinov and Boiadjiev, 1997). The number of iterations was 30 according to the industrial standard ISO 9283 (1998). The trajectories were the same for each iteration with the joint velocities 0.958 rad/s for the first joint and 1.57 rad/s for the second joint and load 10 kg in the end-effector (Dimov, Dobrinov and Boiadjiev, 1997).

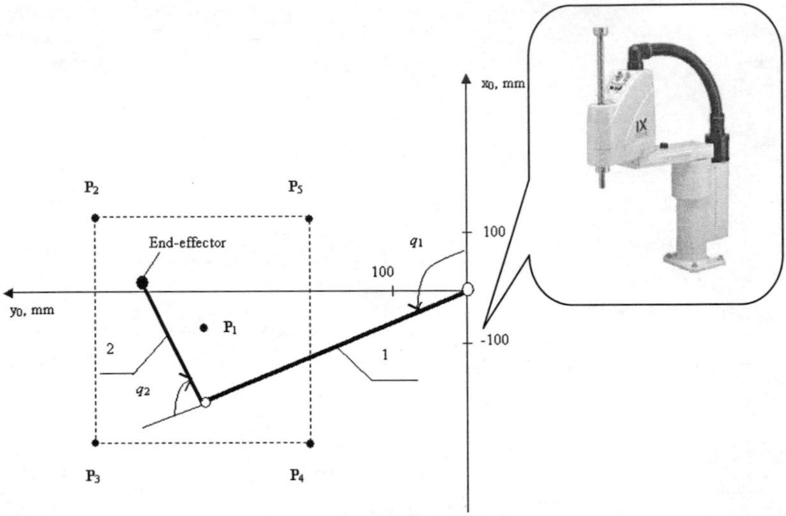

Fig. 2.3 The scheme of part of robot RM 10-01 workspace where positioning accuracy and repeatability measurements were performed (Dimov, Dobrinov and Boiadjiev, 1997)

The method of robot control was "point to point". The control system was realized based on the universal controller 84 EA (Dimov, Dobrinov and Boiadjiev, 1997). The translational joint was fixed in the lowest vertical position ($z = 0$).

In the final points, measurements of positioning accuracy ΔL and positioning repeatability r were performed (the error of the measuring equipment was 1% of the robot end-effector positioning error, which means that at the deviation of 0.1 mm of robot end-effector, the error of the measuring equipment was 0.001 mm) (Dimov, Dobrinov and Boiadjiev, 1997).

Taking into account that Δz component is equal to zero for planar 2-R robotic manipulator, the end-effector positioning accuracy was defined as (ISO 9283, 1998):

$$\Delta L = \sqrt{(\underline{x} - x_c)^2 + (\underline{y} - y_c)^2}, \qquad (2.1)$$

where $\underline{x} = \frac{1}{n}\sum_{i=1}^{n} x_i$ and $\underline{y} = \frac{1}{n}\sum_{i=1}^{n} y_i$ are average values of the Cartesian coordinates of the attained point after n iterations; x_c and y_c are coordinates of the commanded (reference) point in the robot workspace; x_i and y_i are Cartesian coordinates of the attained points in the robot workspace obtained at iteration i and n is the number of iterations.

Table 2.1

Coordinates of points in robot RM 10-01 workspace (coordinate $z = 0$)

Point	Coordinates of points in the base coordinate system		Robot joint coordinates	
	x [mm]	y [mm]	q_1 [rad]	q_2 [rad]
$\mathbf{P}_1$	-80	380	1.17	1.79
$\mathbf{P}_2$	120	580	1.06	0.9
$\mathbf{P}_3$	-280	580	1.77	0.79
$\mathbf{P}_4$	-280	180	2.11	2.236
$\mathbf{P}_5$	120	180	0.34	2.215

Robot end-effector positioning repeatability is defined as (ISO 9283, 1998):

$$r = \underline{D} + 3S_D, \tag{2.2}$$

where $\underline{D}$ is an average deviation after positioning into the command point:

$$\underline{D} = \frac{1}{n}\sum_{i=1}^{n} D_i,$$

D_i is a deviation after positioning into the command point:

$$D_i = \sqrt{(x_i - \underline{x})^2 + (y_i - \underline{y})^2},$$

and S_D is a standard estimation of deviation after positioning into the command point according to the industrial standard ISO 9283 (1998):

$$S_D = \frac{1}{n-1}\sqrt{\sum_{i=1}^{n} (D_i - \underline{D})^2}.$$

The results of experiments are presented in Table 2.2 (Dimov, Dobrinov and Boiadjiev, 1997). The main conclusion that can be formulated based on the experimental results is that end-effector positioning accuracy and repeatability differ significantly in various locations in the robot workspace. For example, end-effector positioning accuracy ΔL in point $\mathbf{P}_4$ (see Table 2.2) was more than five times higher than that in point $\mathbf{P}_1$. In all cases, end-effector positioning accuracy was within the allowed limits (20 μm is a nominal value for industrial robot RM 10-01 (Dimov, Dobrinov and Boiadjiev, 1997)).

Table 2.2

Values of end-effector positioning accuracy and positioning repeatability in five points of robot workspace after 30 iterations (Dimov, Dobrinov and Boiadjiev, 1997)

Points	$\mathbf{P}_1$	$\mathbf{P}_2$	$\mathbf{P}_3$	$\mathbf{P}_4$	$\mathbf{P}_5$
ΔL [μm]	10.9	5.9	1.87	1.73	6.15
r [μm]	6.6	4.6	10.7	2.2	7

We can now go beyond author's conclusions in (Dimov, Dobrinov and Boiadjiev, 1997) and formulate the following additional comments and suggestions. The presented experimental results (see Table 2.2) show that it is highly desirable to conduct pre-investigation of robot end-effector positioning accuracy in the robot workspace before implementing a given robotic system for a given technological operation. For example, in the above-presented case, high-precision operations could be recommended to be performed in point $\mathbf{P}_4$ where both end-effector positioning accuracy and positioning repeatability were the best for the given trajectory. This would allow performing technological operations with a precision that is almost ten times higher than the nominal positioning accuracy (20 μm) for the given industrial robot RM 10-01.

Another important aspect in the investigation of robot end-effector positioning accuracy is how the positioning accuracy changes with the change of the end-effector trajectory. The experimental data obtained by Kieffer, Cahill and James (1997) showed that joint errors for executing similar trajectories in most cases changed less than by 1 % for similar SCARA type 2-R robotic manipulator under computed torque control. However, the large change of the trajectory would change joint error values significantly (exact values depend on the given robot). This assumption is also theoretically proved by the linear dynamic system theory for robots under computed torque control (Kieffer, Cahill and James, 1997). This means that the possibility of robot end-effector positioning accuracy improvement based on the analysis of robot positioning accuracy usually exists only for the same trajectories as those executed at the initial measurements. Therefore, if the robot end-effector trajectory for the given technological operation is changed, new investigations of the end-effector positioning are required. This is a major drawback here, because, in most cases, robotic systems are used for a number of different technological operations requiring an execution of different trajectories in different areas of robot workspace in the working cycle. Despite this fact, based on the presented experimental results of (Dimov, Dobrinov and Boiadjiev, 1997), we still propose to use our Algorithm 2.1 whenever it is possible in order to improve end-effector positioning accuracy for the given technological operation.

Algorithm 2.1

1. Determine allowed areas in robot workspace for performing given technological operation.
2. Plan robotic manipulator end-effector trajectory into the selected points in the allowed area of robot workspace using one of trajectory planning methods (*i.e.*, optimal robot path planning using minimum time criterion (Bobrow, 1988), torque–limited path followed by online trajectory time scaling (Dahl and Nielsen, 1990), torque-optimized trajectory planning (Hollerbach and Suh, 1987), *etc.*).

3. Measure robot end-effector positioning accuracy and repeatability in the selected points of robot workspace while executing planned trajectories in test mode on the available test equipment.
4. Compare the results of measurements of accuracy characteristics in different points and choose a part of robot workspace where end-effector accuracy characteristics were the best.
5. Configure the technological equipment (fixation devices, assembly devices, *etc.*) on the working table in the chosen part of robot workspace with the best end-effector positioning accuracy characteristics based on performed tests and carry out given technological operations by robot there.

As it is mentioned in Algorithm 2.1, first, one should determine allowed areas in the robot workspace for performing the given technological operation. This means that it is important to consider different options for performing given technological operations in the robot workspace. After planning time optimal robotic manipulator end-effector trajectories into the identified allowed parts of robot workspace, these options should be experimentally investigated to determine end-effector positioning accuracy characteristics. If the special measuring equipment for measuring end-effector positioning accuracy is not available, then one can measure joint errors based on the embedded positioning sensors in manipulator joints and use manipulator kinematic function to calculate end-effector positioning accuracy characteristics. After analysis of all measurement results, one may choose those parts of the robot workspace where robot end-effector accuracy characteristics were the best for performing the given technological operation. The last stage usually includes the configuration of technological equipment (fixation devices, assembly devices, *etc.*) in the chosen part of the robot workspace with the best end-effector positioning accuracy characteristics.

The implementation of Algorithm 2.1 in practice allows an optimal robot accuracy usage for the given model of industrial robot in the given industrial application. However, as it was mentioned earlier, such approach has an important limitation for its usage, because in practice one usually has to use the whole workspace of the robotic manipulator. Therefore, Algorithm 2.1 is difficult to apply in practice, due to the fact that this would limit the workspace of the robotic manipulator.

In order to improve end-effector positioning accuracy, we propose to use joint error maximum mutual compensation, which is described in the next section and is based on the non-uniformity of the robot positioning accuracy characteristics in the robot workspace.

2.2. JOINT ERROR MAXIMUM MUTUAL COMPENSATION IN NONREDUNDANT ROBOTIC MANIPULATORS

In practice, one needs to use the whole workspace of the robotic manipulator. For example, when a robot performs some technological operation along the conveyor, it is impossible to change the part of the robot workspace. However, it is still often possible to change the location of the working point at some small value (for example, up to 10 mm).

To benefit from the non-uniformity of the robot positioning accuracy characteristics (see Chapter 2.1) in the robot workspace, we propose to use a slight change of the end-effector working position that allows performing local optimization of the end-effector positioning accuracy by using joint error maximum mutual compensation. We will further present our approach based on the 2-R planar robotic manipulator.

For the kinematic scheme of 2-R robotic manipulator shown in Fig. 2.4 (1 and 2 are respectively, first and second links of the robotic manipulator; x_0y_0 is the base robot coordinate system; x_1y_1 and x_2y_2 are, accordingly, the Cartesian coordinate systems of first and second links; q_1 and q_2 are first and second robot joint coordinates), elementary end-effector Cartesian errors Δx and Δy in the base coordinate system can be defined as (Smolnikov, 1990):

$$\Delta x = x - x_c$$

$$\Delta y = y - y_c \, ,$$

where x, y are Cartesian coordinates of the attained pose and x_c, y_c are the command pose coordinates. Cartesian coordinates x, y and x_c, y_c can be expressed as (Smolnikov, 1990):

$$x = l_1 \cos q_1 + l_2 \cos (q_1 + q_2)$$

$$y = l_1 \sin q_1 + l_2 \sin (q_1 + q_2)$$

$$x_c = l_1 \cos (q_1 + \Delta q_1) + l_2 \cos (q_1 + \Delta q_1 + q_2 + \Delta q_2)$$

$$y_c = l_1 \sin (q_1 + \Delta q_1) + l_2 \sin (q_1 + \Delta q_1 + q_2 + \Delta q_2).$$

After this, errors Δx and Δy can be expressed as:

$$\Delta x = l_1 \cos q_1 + l_2 \cos (q_1 + q_2) - l_1 \cos (q_1 + \Delta q_1) - l_2 \cos (q_1 + \Delta q_1 + q_2 + \Delta q_2)$$

$$\Delta y = l_1 \sin q_1 + l_2 \sin (q_1 + q_2) - l_1 \sin (q_1 + \Delta q_1) - l_2 \sin (q_1 + \Delta q_1 + q_2 + \Delta q_2).$$

and after further algebraic transformations as:

$$\begin{aligned} \Delta x &= l_1 \cos q_1 + l_2 \cos(q_1+q_2) - l_1(\Delta q_1 \sin q_1 + \cos q_1) - l_2[\cos (q_1+q_2) + (\Delta q_1 + \Delta q_2) \sin (q_1+q_2)] \\ &= -(l_1 \sin q_1 + l_2 \sin (q_1 + q_2))\, \Delta q_1 - l_2 \sin (q_1 + q_2)\, \Delta q_2 \end{aligned}$$

(2.3)

$$\begin{aligned} \Delta y &= l_1 \sin q_1 + l_2 \sin (q_1+q_2) - l_1 (\sin q_1 + \Delta q_1 \cos q_1) - l_2 [\sin(q_1+q_2) + (\Delta q_1 + \Delta q_2) \cos(q_1+q_2)] \\ &= (l_1 \cos q_1 + l_2 \cos (q_1 + q_2))\, \Delta q_1 + l_2 \cos (q_1 + q_2)\, \Delta q_2 \, , \end{aligned}$$

where q_1 and q_2 are joint coordinates, Δq_1 is an elementary joint error of the first joint, Δq_2 is an elementary joint error of the second joint; l_1 and l_2 are, accordingly, the lengths of manipulator links.

The square of the Cartesian coordinate end-effector error ΔL^2 for 2-R robotic manipulator can be found as:

$$\Delta L^2 = \Delta x^2 + \Delta y^2,$$

and since

$$x = l_1 \cos q_1 + l_2 \cos (q_1 + q_2)$$

$$y = l_1 \sin q_1 + l_2 \sin (q_1 + q_2),$$

using (2.3), one can express Δx and Δy as:

$$\Delta x = -y\, \Delta q_1 - l_2 \sin (q_1 + q_2)\, \Delta q_2$$

$$\Delta y = x\, \Delta q_1 + l_2 \cos (q_1 + q_2)\, \Delta q_2,$$

and then after a few algebraic transformations find ΔL^2 as:

$$\Delta L^2 = \Delta q_1^2 (l_1^2 + l_2^2) + 2\, l_1 l_2 (\Delta q_1^2 + \Delta q_1 \Delta q_2) \cos q_2 + l_2^2\, \Delta q_2^2 + 2 \Delta q_1\, \Delta q_2\, l_2^2, \qquad (2.4)$$

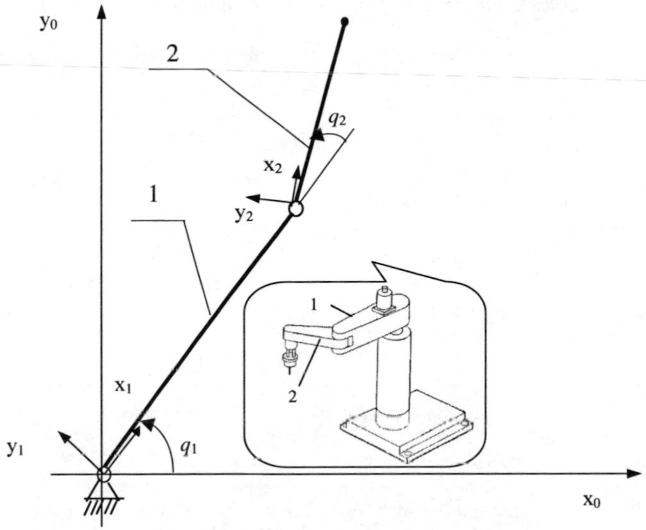

Fig. 2.4. The kinematic scheme of 2-R robotic manipulator

As one can see from (2.4), we have got an important result, namely, ΔL does not depend on joint coordinate q_1. The best end-effector positioning accuracy for 2-R robotic manipulator can be obtained when $\Delta L^2 = 0$ in (2.4). In this case, the optimal joint coordinate q_{2opt} for which the joint errors Δq_1 and Δq_2 will be maximally or fully mutually compensated can be found by taking a derivative of ΔL^2 (2.4) with respect to q_2. The derivative $\frac{\partial(\Delta L^2)}{\partial q_2}$ is:

$$\frac{\partial(\Delta L^2)}{\partial q_2} = -\, 2\, l_1 l_2 (\Delta q_1^2 + \Delta q_1 \Delta q_2) \sin q_2\, . \qquad (2.5)$$

Now, we just have to investigate (2.5) on maximum and minimum by equalizing it to zero:

$$-2\, l_1 l_2(\Delta q_1^2 + \Delta q_1 \Delta q_2) \sin q_2 = 0.$$

After equalizing (2.5) to zero, we can find the following two solutions (critical points): $q_2 = 0°$ and $q_2 = 180°$ (solutions like $q_2 = 360°$, $q_2 = 540°$ *etc.* are not taken into account because they repeat solutions $q_2 = 0°$ or $q_2 = 180°$), which could be a maximum or minimum for (2.4). To find out which of these values ($q_2 = 0°$ or $q_2 = 180°$) is maximum and minimum we will have to carry out additional investigations.

Values $q_2 = 0°$ and $q_2 = 180°$ are relatively seldom used in robotic manipulators, thus, we use the obtained solutions as ideal ones in our further discussions. It is now important to identify which of the values ($q_2 = 0°$ or $q_2 = 180°$) gives a minimum and which gives a maximum. One can find this out if one would attempt to input these values into (2.4) with different combinations of signs of q_2, Δq_1 and Δq_2 and observe the value of ΔL^2 on maximum and minimum.

As one can see from (2.4), the only component of the equation that is dependent on the signs of q_2, Δq_1 and Δq_2 is:

$$2\, l_1 l_2(\Delta q_1^2 + \Delta q_1 \Delta q_2) \cos q_2\, .$$

Other components from (2.4) have always positive signs. Thus, we will use only this component in our investigation on maximum and minimum of (2.4). If the sign of this component is negative with the given values of Δq_1, Δq_2, l_1, l_2 and q_2 (0° or 180°), then the given value ($q_2 = 0°$ or $q_2 = 180°$) gives a minimum of ΔL^2 in (2.4). The investigation on maximum and minimum (the calculation of ΔL^2 based on various combinations of values and signs of q_2, Δq_1 and Δq_2 as inputs in (2.4)) for 2-R planar robotic manipulator with the link lengths of $l_1 = 0.4$ m and $l_2 = 0.25$ m is presented in Table 2.3.

The results presented in Table 2.3 show that $q_2 = 0°$ is a minimum for (2.4) only if $|\Delta q_2| > |\Delta q_1|$ and the signs of Δq_2 and Δq_1 are opposite. In all other cases $q_2 = 0°$ is a maximum (see Table 2.3). The value of $q_2 = 180°$ is minimum for (2.4) in all cases except for those in which $q_2 = 0°$ is a minimum (see Table 2.3). The link lengths do not influence the sign of component $(2\, l_1 l_2(\Delta q_1^2 + \Delta q_1 \Delta q_2) \cos q_2)$ from (2.4) and are simply taken from robot RM 10-01, as an example. Table 2.3 can be used for any 2-R SCARA type robotic manipulator.

Based on the results presented in Table 2.3, we can formulate the following conclusion. If we suppose that joint error values and signs exist based on the law of equal probability (it is normally the case for random values, as those in normal Gaussian distribution), one does not necessarily have to know joint errors to perform joint error maximum mutual compensation. For example, if joint error values are not known to perform joint error maximum mutual compensation then, in most cases, with the probability $p_{improv} = 0.75$ (to be explained further in the text) of end-effector positioning accuracy improvement, the value of $q_2 = 180°$ will be optimal for joint error maximum mutual compensation in 2-R robotic manipulators. The probability of end-effector positioning accuracy improvement $p_{improv} = 0.75$ has been calculated taking into account that the probability of $|\Delta q_2| > |\Delta q_1|$ is equal to 0.5 (another only possible event that $|\Delta q_2| \leq |\Delta q_1|$ has also probability of 0.5) and the probability that signs of Δq_2 and Δq_1 are opposite is equal to 0.5 (another only possible event that the signs are the same has also probability of 0.5). All these probability values are found based on the law of equal probability (Gmurman, 2000). The probability of both events and, thus, the probability of end-effector positioning accuracy deterioration $p_{deterior}$ when $q_2 = 180°$, is a multiplication:

$$p_{deterior} = p(|\Delta q_2| > |\Delta q_1|)\, p(\text{sign}\Delta q_2 \neq \text{sign}\Delta q_1) = 0.5\ 0.5 = 0.25.$$

where $p(|\Delta q_2| > |\Delta q_1|)$ is the probability that $|\Delta q_2| > |\Delta q_1|$ and $p(\text{sign}\Delta q_2 \neq \text{sign}\Delta q_1)$ is the probability that signs of Δq_1 and Δq_2 are different. The probability that the end-effector positioning accuracy deterioration does not happen, in its turn, can be found as (Gmurman, 2000):

$$p_{improv} = 1 - p_{deterior} = 1 - 0.25 = 0.75.$$

where p_{improv} is the probability that end-effector positioning accuracy will be improved when $q_2 = 180°$.

Table 2.3

Investigation on minimum and maximum of end-effector positioning accuracy ΔL for 2-R planar robotic manipulator with link lengths $l_1 = 0.4$ m and $l_2 = 0.25$ m

№	Δq_1 [rad]×10^{-6}	Δq_2 [rad]×10^{-6}	Relation	q_2 [rad]	ΔL [μm]	Definition
1	2.29	2.11	$\lvert\Delta q_2\rvert \leq \lvert\Delta q_1\rvert$ &	0	1.500	max
			$\text{sign}\Delta q_2 = \text{sign}\Delta q_1$	π	1.359	min
2	2.29	-2.11	$\lvert\Delta q_2\rvert \leq \lvert\Delta q_1\rvert$ &	0	0.921	max
			$\text{sign}\Delta q_2 \neq \text{sign}\Delta q_1$	π	0.912	min
3	-2.29	2.11	$\lvert\Delta q_2\rvert \leq \lvert\Delta q_1\rvert$ &	0	0.921	max
			$\text{sign}\Delta q_2 \neq \text{sign}\Delta q_1$	π	0.912	min
4	-2.29	-2.11	$\lvert\Delta q_2\rvert \leq \lvert\Delta q_1\rvert$ &	0	1.500	max
			$\text{sign}\Delta q_2 = \text{sign}\Delta q_1$	π	1.359	min
5	2.11	2.29	$\lvert\Delta q_2\rvert \geq \lvert\Delta q_1\rvert$ &	0	1.452	max
			$\text{sign}\Delta q_2 = \text{sign}\Delta q_1$	π	1.318	min
6	2.11	-2.29	$\lvert\Delta q_2\rvert > \lvert\Delta q_1\rvert$ &	0	0.841	**min**
			$\text{sign}\Delta q_2 \neq \text{sign}\Delta q_1$	π	0.850	**max**
7	-2.11	2.29	$\lvert\Delta q_2\rvert > \lvert\Delta q_1\rvert$ &	0	0.841	**min**
			$\text{sign}\Delta q_2 \neq \text{sign}\Delta q_1$	π	0.850	**max**
8	-2.11	-2.29	$\lvert\Delta q_2\rvert \geq \lvert\Delta q_1\rvert$ &	0	1.452	max
			$\text{sign}\Delta q_2 = \text{sign}\Delta q_1$	π	1.318	min

The analysis of the above-presented investigation shows that 2-R planar robotic manipulator with given values of joint errors Δq_1 and Δq_2 has a special configuration with q_2

= 180° when the value of end-effector positioning accuracy ΔL is, in most cases (see Table 2.3), the best, because of the maximum mutual compensation of joint errors in such configuration. One can use this effect in practice. Let us define a term "joint error maximum mutual compensation" and use it when one attempts to find the configuration of robotic manipulator in which the best end-effector positioning accuracy under the given values of joint errors Δq_1 and Δq_2 exists.

There are basically three use cases for joint error maximum mutual compensation in practice:

1. Joint error maximum mutual compensation can be applied when joint error values for a given robotic manipulator are known. In this case, one can directly use formula (2.4) and (2.5) and find q_{2opt} for which joint errors Δq_1 and Δq_2 will be maximally or fully mutually compensated.
2. Joint error maximum mutual compensation can be applied when joint error values for a given robotic manipulator are not known. In this case, one can with relatively high probability of 0.75 improve end-effector positioning accuracy. This is the most important result of the given research work, because normally one needs to know or approximate joint errors prior to their compensation using, for example, typical methods of correcting planned trajectories with the known values of joint errors (*e.g.*, robot calibration). Hence, in case of unknown joint errors, one can take the value of q_2 = 180° as the optimal one and with the probability 0.75 of end-effector positioning accuracy improvement use joint error maximum mutual compensation. This means there is also a probability of 0.25 that an attempt to use joint error maximum mutual compensation will lead to the opposite effect, namely, to a decrease of the end-effector positioning accuracy. As one can see it from Table 2.3, the maximum possible end-effector positioning accuracy improvement obtained between maximum and minimum of end-effector positioning accuracy was ten percent (see rows 1 and 4 in Table 2.3). Ten percent improvement of end-effector positioning accuracy without investment in expensive high-precision hardware is not a bad result. The maximum possible deterioration obtained between maximum and minimum of end-effector positioning accuracy was only one percent (see rows 6 and 7 in Table 2.3). This means that the possible negative effect in case of end-effector positioning accuracy deterioration for q_2 = 180° is very small.
3. Joint error maximum mutual compensation can be applied when some joint error values for a given robotic manipulator are known and some are not (for example, Δq_1 is known and Δq_2 is not known or the opposite combination). In this case, one will have the same situation as it is described above for item 2, namely, one can take the value of q_2 = 180° as the optimal one and with the probability 0.75 of end-effector positioning accuracy improvement use joint error maximum mutual compensation.

The dependences of the end-effector positioning accuracy of 2-R robotic manipulator, for example, for robot RM 10-01 with different link lengths l_1 and l_2, as well as different permanent joint errors Δq_1 and Δq_2, based on (2.4), are shown in Fig. 2.5 – 2.8. One can see from Fig. 2.5 – 2.8 that an increase of link lengths (l_1 or l_2) and/or joint error absolute values (Δq_1 or Δq_2) increases the amplitude of the *cos* curve and, thus, end-effector positioning accuracy becomes worse (ΔL increases).

Let us state a problem more precisely based on the available results presented in this chapter so far. In practice, optimal angle value q_{2opt} = 180°, where joint error maximum mutual compensation takes place (see equation (2.5)), is not applicable, because of difficulties to use robotic manipulators for performing given technological operations in that configuration. Additionally, if one would attempt to approach the optimal joint coordinate by changing joint coordinates in 2-R planar robotic manipulator, joint errors Δq_1 and Δq_2 would change as well (Kieffer, Cahill and James, 1997). As it follows from the experimental results presented in (Kieffer, Cahill and James, 1997), average values of joint errors Δq_1 and Δq_2 will change significantly in the working point with the optimal joint coordinate q_{2opt} if it is far from the initial value q_2. Experimental data from (Kieffer, Cahill and James, 1997) shows that average values of joint errors Δq_1 and Δq_2 will be the same only for similar trajectories in the given area of the robot workspace.

$\Delta q_1 = -2.11 \times 10^{-6}$ rad
$\Delta q_2 = -2.29 \times 10^{-6}$ rad
$l_2 = 0.25$ m

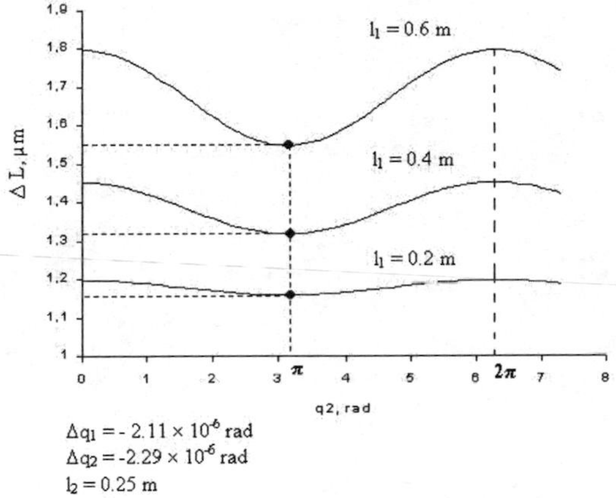

Fig. 2.5. Graph of dependence of robot end-effector positioning accuracy ΔL on the distance from the optimal joint coordinate $q_{2opt} = 180°$ with various l_1

$\Delta q_1 = -2.11 \times 10^{-6}$ rad
$\Delta q_2 = -2.29 \times 10^{-6}$ rad
$l_1 = 0.4$ m

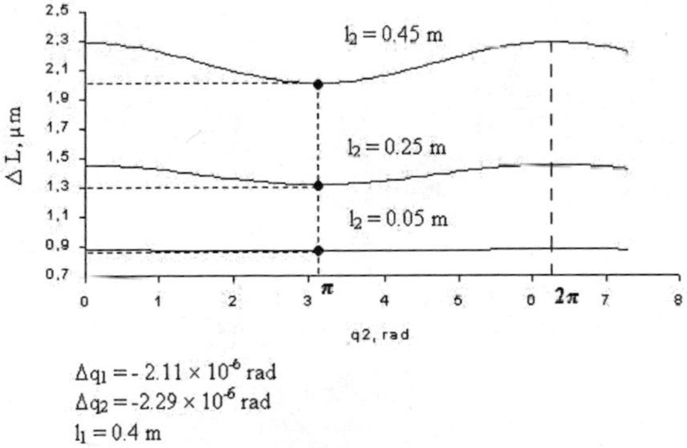

Fig. 2.6. Graph of dependence of robot end-effector positioning accuracy ΔL on the distance from the optimal joint coordinate $q_{2opt} = 180°$ with various l_2

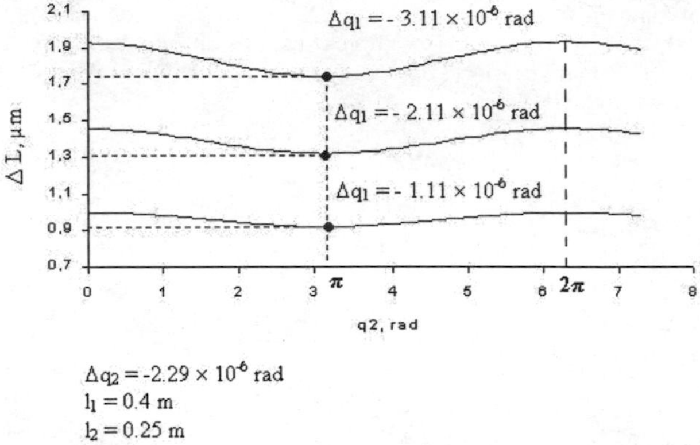

Fig. 2.7. Graph of dependence of robot end-effector positioning accuracy ΔL on the distance from the optimal joint coordinate $q_{2opt} = 180°$ with various Δq_1

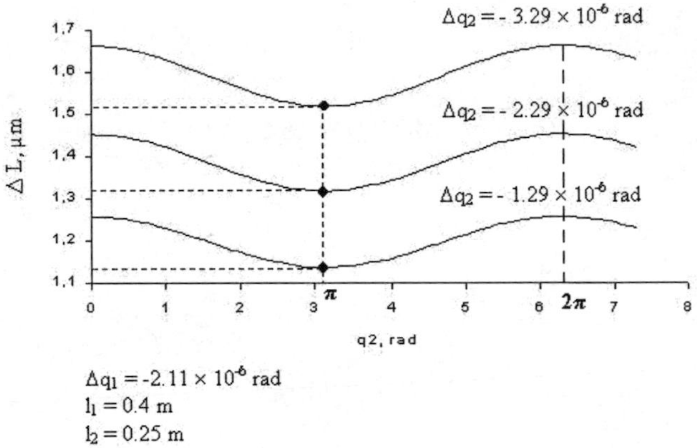

Fig. 2.8. Graph of dependence of robot end-effector positioning accuracy ΔL on the distance from the optimal joint coordinate $q_{2opt} = 180°$ with various Δq_2

We propose to use joint error mutual compensation and improve robot end-effector positioning accuracy ΔL by performing a local optimization (local means within a limited working area of robot) of the end-effector positioning accuracy, in particular, by changing the position of initial working point $\mathbf{P}_{initial}$ into $\mathbf{P}_{final}$, as it is shown in Fig. 2.9. Fig. 2.9 shows the dependence of the square of robot end-effector positioning accuracy ΔL^2 on the value of second joint coordinate q_2 for robot RM 10-01 with link lengths $l_1 = 0.4$ m, $l_2 = 0.25$ m and joint errors Δq_1 = -2.11 $\times 10^{-6}$ rad, Δq_2 = -2.29 $\times 10^{-6}$ rad. One can see from Fig. 2.9 that the closer q_2 value to the optimal joint coordinate q_{2opt} = 180°, where joint error maximum mutual compensation takes place, the better end-effector positioning accuracy is. Thus, for example, by moving along the curve (see Fig. 2.9), end-effector positioning accuracy in $\mathbf{P}_{final}$ is better than it is in $\mathbf{P}_{initial}$. Fig. 2.10 shows how the change of $\mathbf{P}_{initial}$ to $\mathbf{P}_{final}$ will look like for 2-R robotic manipulator. One can see the new robotic manipulator configuration (presented with the dashed line) for $\mathbf{P}_{final}$ in Fig. 2.10. In order to get into $\mathbf{P}_{final}$, one has to change coordinate q_2 of the given initial working point $\mathbf{P}_{initial}$ in the direction of the q_{2opt} = 180° on some value δq_2 (δq_1, δq_2 and a_{dist} are shown in Fig. 2.10 and will be later explained in this chapter), so that the value of joint coordinate q_2 becomes as close as possible to the joint coordinate q_{2opt} (see Fig 2.7, the closer q_2 to q_{2opt} the better end-effector positioning accuracy is obtained under the same joint errors).

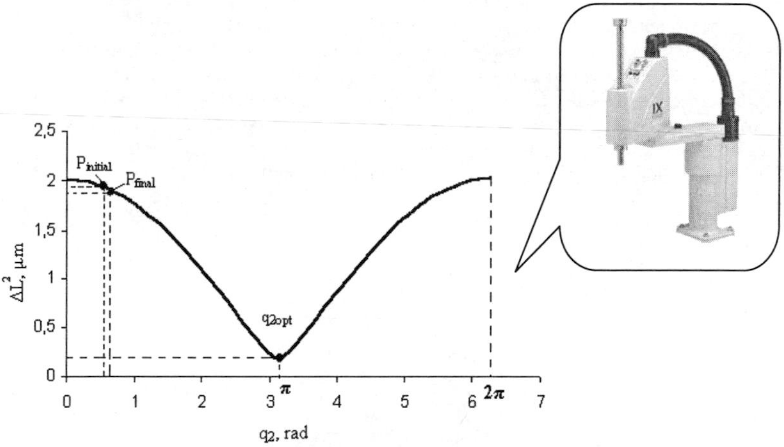

Fig. 2.9. Graph of dependence of square of robot end-effector positioning accuracy ΔL^2 on second joint coordinate q_2 (based on equation (2.4))

It is supposed here, based on (Kieffer, Cahill and James, 1997), that the change of the initial working point $\mathbf{P}_{initial}$ into point $\mathbf{P}_{final}$ will not change average values of joint errors significantly because of the relatively small change (not more than 10 mm) of the end-effector position. As the result of the change of the working point from $\mathbf{P}_{initial}$ to $\mathbf{P}_{final}$, one can expect better joint error mutual compensation in point $\mathbf{P}_{final}$ with new q_2 and, as a result, robot end-effector positioning accuracy will be improved (see Fig. 2.9).

A slight change δq_1 of first joint coordinate q_1 in Fig. 2.10 is needed only to have the value of δq_2 as big as possible, so that the final end-effector position remains within the distance of a_{dist} from its initial position. This means that without a slight change δq_1 of first joint coordinate q_1 one would not be able to have a maximum possible value of δq_2, so that the final end-effector position is within the distance of a_{dist}.

To show how the above-presented scheme will work in practice, an example for 2-R planar robotic manipulator is presented below. One can consider the allowed change a_{dist} = 10 mm of the initial working point $\mathbf{P}_{initial}$ to perform a given technological operation. It is supposed that with the change of $\mathbf{P}_{initial}$ location within a_{dist} = 10 mm, average values of joint errors Δq_1 and Δq_2 will not change (in practice, they may change slightly (Kieffer, Cahill and James, 1997) and these changes will largely depend on the given robotic manipulator). Let us suppose that a given technological operation must be executed in point $\mathbf{P}_{initial}$ (see Fig. 2.10) by the robotic manipulator RM 10-01 with the same parameters of the trajectory as those in the experiments from (Dimov, Dobrinov and Boiadjiev, 1997) for point $\mathbf{P}_3$ (see Fig. 2.3) with $q_1 = 1.77$ rad and $q_2 = 0.79$ rad. The end-effector positioning accuracy was $\Delta L = 1.87$ μm in point $\mathbf{P}_{initial}$. The joint error values $\Delta q_1 = 2.29 \times 10^{-6}$ rad and $\Delta q_2 = 2.11 \times 10^{-6}$ rad were found in point $\mathbf{P}_{initial}$ based on the manipulator kinematic model. The optimal value of the joint coordinate $q_{2opt} = 180°$ was found for the given conditions using (2.5) (see also Fig. 2.9). According to the previous discussions, the value of the initial joint coordinate $q_2 = 0.79$ rad in point $\mathbf{P}_{initial}$ should be increased at some value to become closer to the optimal value $q_{2opt} = 180°$. Assuming that the allowed distance from the $\mathbf{P}_{initial}$ to a new $\mathbf{P}_{final}$ may not be more than 10 mm, one can increase the second joint coordinate q_2 by $\delta q_2 = 0.08$ rad (see Fig. 2.10) and the first joint coordinate should be decreased by $\delta q_1 = -0.01$ rad (see Fig. 2.10) to get the maximum mutual joint error compensation within the allowed region of 10 mm. The change of q_1 by $\delta q_1 = -0.01$ rad and q_2 by $\delta q_2 = 0.08$ rad, accordingly, for the first joint and second joints have been analytically found using triangles formed by manipulator links (see Fig. 2.10 and 2.11) and based on the assumption that the resulting position of the end-effector $\mathbf{P}_{final}$ will not exceed the maximum allowed value of $a_{dist} = 10$ mm from the initial point $\mathbf{P}_{initial}$. How the values of δq_1 and δq_2 can be found for 2-R planar robotic manipulator will be further explained in this chapter.

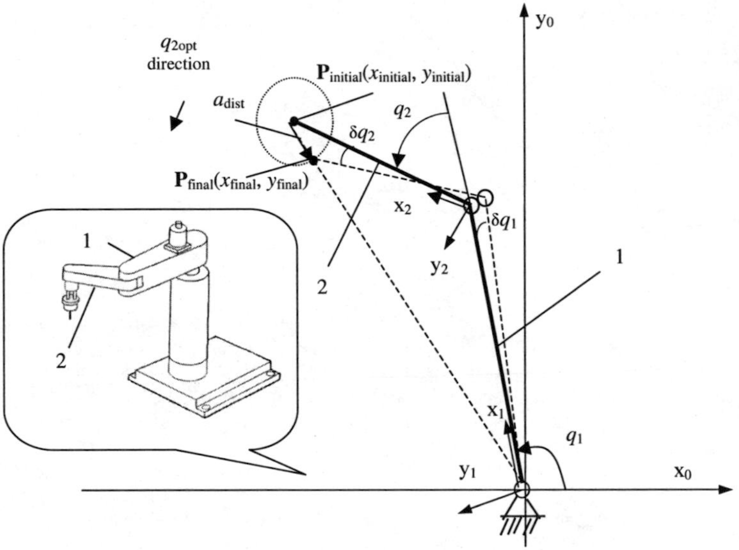

Fig. 2.10. The kinematic scheme of 2-R robotic manipulator and local optimization of robot end-effector positioning accuracy

After the above-mentioned changes of joint coordinates using δq_1 and δq_2, as it is shown in Fig. 2.10, the new joint coordinates in point $\mathbf{P}_{final}$ became $q_1 = 1.76$ rad and $q_2 = 0.87$ rad.

Taking into account that the old values of joint errors did not change by more than 1% (Kieffer, Cahill and James, 1997), one can use the old values of joint errors Δq_1 and Δq_2 in order to determine robot end-effector positioning accuracy. Based on (2.4), the new value of robot end-effector positioning accuracy $\Delta L = 1.81$ μm was found. The comparison of the new end-effector positioning accuracy $\Delta L = 1.81$ μm in point $\mathbf{P}_{\text{final}}$ with old one $\Delta L = 1.87$ μm in point $\mathbf{P}_{\text{initial}}$ (see value of end-effector positioning accuracy in point $\mathbf{P}_3$ in Table 2.2) shows that the end-effector positioning accuracy improved by 3.2 % due to the better joint error compensation in the new point $\mathbf{P}_{\text{final}}$.

The simplified picture of 2-R robotic manipulator is shown in Fig. 2.11 to explain how the values of δq_1 and δq_2 can be found.

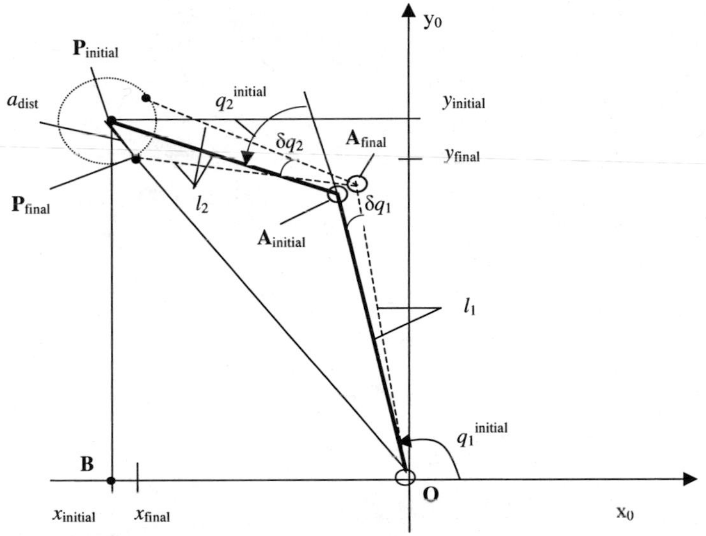

Fig. 2.11. Geometry of 2-R robotic manipulator with end-effector located in points $\mathbf{P}_{\text{initial}}$ and $\mathbf{P}_{\text{final}}$ (dashed lines were used to present robotic manipulator configuration in $\mathbf{P}_{\text{final}}$)

As one can see in Fig. 2.11, point $\mathbf{P}_{\text{initial}}$ is in the center of the circle with the radius equal to a_{dist}. Within this circle with radius a_{dist}, point $\mathbf{P}_{\text{final}}$ with the better joint error mutual compensation than in $\mathbf{P}_{\text{initial}}$ can be found. As discussed previously, joint error maximum mutual compensation would take place if $q_{2\text{opt}} = 180°$ rad. In the given example, it is not possible to realize $q_2 = 180°$ rad because we have a limitation, namely, the end effector position change should not exceed a_{dist}. The graph in Fig. 2.9 shows that the closer q_2 to $q_{2\text{opt}} = 180°$ rad the better joint error mutual compensation is. Thus, one can find the location of point $\mathbf{P}_{\text{final}}$ (see Fig. 2.11) if one looks for a point within a_{dist} from $\mathbf{P}_{\text{initial}}$ that has the largest value q_2 (*i.e.*, in this case, the larger the value of q_2 is the closer it is to $q_{2\text{opt}} = 180°$ rad and, thus, the better joint error mutual compensation is). One can find $\mathbf{P}_{\text{final}}$ analytically if one connects $\mathbf{P}_{\text{initial}}$ with point **O** (located in the center of base coordinates) and takes into account that (q_2 + angle $PAO = 180°$). Hence, the smaller the angle PAO is the larger q_2 is ($\mathbf{A}_{\text{initial}}$ and $\mathbf{A}_{\text{final}}$ are the points located in the center of second joint; **A** is the general alias for the center of second joint and **P** is the general alias for end-effector). Additionally, it is known from geometry of triangles that the smaller the length of one of the triangle legs is (if other triangle legs' lengths are not changed) the smaller the opposite angle value is. In our case, this means in order to find the smallest value for angle *PAO* one has to find the smallest possible triangle leg $\mathbf{P}_{\text{final}}\mathbf{O}$.

It is clear from the geometry of triangles and circles (see Fig. 2.11) that the smallest triangle leg $\mathbf{P}_{\text{final}}\mathbf{O}$ can be obtained if $\mathbf{P}_{\text{final}}\mathbf{O} = \mathbf{P}_{\text{initial}}\mathbf{O} - a_{\text{dist}}$ (namely, it is located on the line that includes radius a_{dist} and connects $\mathbf{P}_{\text{initial}}$ and the center of coordinates **O**). As a result, we found analytically the location of $\mathbf{P}_{\text{final}}$, where improved joint error mutual compensation takes place One can find new joint coordinates q_1^{final}, q_2^{final} and the required changes δq_1, δq_2 ($q_1^{\text{final}} = q_1^{\text{initial}} + \delta q_1$; $q_2^{\text{final}} = q_2^{\text{initial}} + \delta q_2$). q_1^{final} (not shown in Fig. 2.11, but can be easily found) can be found using the triangle $\mathbf{P}_{\text{final}}\mathbf{A}_{\text{final}}\mathbf{O}$ (see new manipulator configuration shown with the dashed lines with the end-effector in point $\mathbf{P}_{\text{final}}$ in Fig. 2.11) and with the following conversions:

$$q_1^{\text{final}} = 180 - \text{angle } P_{\text{final}}OB - \text{angle } A_{\text{final}}OB$$

$$= 180 - \arccos\left(\frac{-l_2^2 + l_1^2 + (\sqrt{x_{\text{initial}}^2 + y_{\text{initial}}^2} - a_{\text{dist}})^2}{2l_1(\sqrt{x_{\text{initial}}^2 + y_{\text{initial}}^2} - a_{\text{dist}})}\right) - \left|\text{arctg}\left(\frac{y_{\text{initial}}}{x_{\text{initial}}}\right)\right|.$$

q_2^{final} can be found from

$$x = l_1 \cos q_1 + l_2 \cos(q_1 + q_2).$$

This means that

$$q_2^{\text{final}} = \arccos\left(\frac{x_{\text{final}} - l_1\cos(q_1^{\text{final}})}{l_2}\right) - q_1^{\text{final}} =$$

$$= \arccos\left(\frac{\left(\sqrt{x_{\text{initial}}^2 + y_{\text{initial}}^2} - a_{\text{dist}}\right)\cos\left(\left|\text{arctg}\left(\frac{y_{\text{initial}}}{x_{\text{initial}}}\right)\right|\right) - l_1\cos(q_1^{\text{final}})}{l_2}\right) - q_1^{\text{final}}.$$

After finding q_1^{final} and q_2^{final}, one can easily find rotation values δq_1 and δq_2:

$$\delta q_1 = q_1^{\text{final}} - q_1^{\text{initial}}$$

$$= 180 - \arccos\left(\frac{-l_2^2 + l_1^2 + (\sqrt{x_{\text{initial}}^2 + y_{\text{initial}}^2} - a_{\text{dist}})^2}{2l_1(\sqrt{x_{\text{initial}}^2 + y_{\text{initial}}^2} - a_{\text{dist}})}\right) - \left|\text{arctg}\left(\frac{y_{\text{initial}}}{x_{\text{initial}}}\right)\right| - q_1^{\text{initial}}, \tag{2.6}$$

$$\delta q_2 = q_2^{\text{final}} - q_2^{\text{initial}}$$

$$= \arccos\left(\frac{\left(\sqrt{x_{\text{initial}}^2 + y_{\text{initial}}^2} - a_{\text{dist}}\right)\cos\left(\left|\text{arctg}\left(\frac{y_{\text{initial}}}{x_{\text{initial}}}\right)\right|\right) - l_1\cos(q_1^{\text{final}})}{l_2}\right) - q_1^{\text{final}} - q_2^{\text{initial}}. \tag{2.7}$$

Before applying obtained values of δq_1 and δq_2 to the given robotic manipulator, one has to take into account their signs (*i.e.*, use functions ATAN2, *etc.* in real robot applications). One should also take into account the signs of q_1 and q_2 during calculation of δq_1 and δq_2 in (2.6) and (2.7). The rotation values δq_1 and δq_2 could be also easily found using computer simulation in MatCAD, MatLab, Microsoft Excel or other mathematical software packages by simply iterating all possible end-effector coordinates within the distance a_{dist} from the initial working point and finding the best value of end-effector positioning accuracy using (2.4).

To summarize the obtained results, let us formulate the method of joint error maximum mutual compensation as Algorithm 2.2, which is valid for 2-R plane robotic manipulator and can be used to improve robot end-effector positioning accuracy ΔL by changing the position of initial working point $\mathbf{P}_{initial}$ into $\mathbf{P}_{final}$, as it is shown in Fig. 2.9.

Algorithm 2.2

1. Determine a part of robot workspace in which the given technological operation is to be carried out.
2. Determine the working point $\mathbf{P}_{initial}$ in the given part of the robot workspace (usually, the choice of this point is limited by the location of technological equipment).
3. Determine the allowed change a_{dist} of the location of the chosen working point $\mathbf{P}_{initial}$.
4. Determine, using (2.4), (2.5), (2.6) and (2.7), a new point $\mathbf{P}_{final}$, which is located within a_{dist} from $\mathbf{P}_{initial}$, in the robot workspace with joint error maximum mutual compensation.
5. Plan a trajectory into point $\mathbf{P}_{final}$ so that the new configuration of the robotic manipulator with the better mutual compensation of joint errors and, thus, improved positioning accuracy, is used for a given technological operation.

The core of the developed method is to take into account joint error maximum mutual compensation when one determines working points in the robot workspace. This includes three typical use cases that can be met in practice. First, one can use joint error maximum mutual compensation to compensate some known or well-predictable joint errors (when it is impossible or inconvenient to compensate them at the control system level). Second, one can use joint error maximum mutual compensation to improve end-effector positioning accuracy by mutually compensating unknown joint errors and, thus, providing end-effector positioning accuracy improvement. If joint error values are not known then with the probability of 0.75 the use of joint error maximum mutual compensation for 2-R robotic manipulators will improve end-effector positioning accuracy. Third, one can use joint error maximum mutual compensation to improve end-effector positioning accuracy by mutually compensating partly known joint errors. This will provide end-effector positioning accuracy improvement, which is the same as it was in the second use case (with all unknown joint errors).

All above-mentioned use cases are universal and are valid for all robotic manipulators that contain at least two rotary joints in one plane. Therefore, the local improvement of the end-effector positioning accuracy can be similarly performed for other types of robots with more than one pair of two joints in one plane. In this case, one should improve end-effector positioning accuracy by presenting robotic manipulators as chains of 2-R joints and applying the developed method to those chains (see Chapter 2.4 for further details). The implementation of the developed method in practice can increase the accuracy of the given technological operations in the given robot workspace by 1 to 15 percent in most cases.

2.3. USE OF JOINT ERROR MAXIMUM MUTUAL COMPENSATION IN INDUSTRIAL ROBOTS

As it was mentioned in Chapter 2.2, one can use joint error maximum mutual compensation to improve end-effector positioning accuracy by compensating unknown joint errors. In this case, an improvement of end-effector positioning accuracy will be obtained for $q_{2opt} = 180°$ if joint error $|\Delta q_2|$ is not higher than $|\Delta q_1|$ and joint error signs are not the same (sign(Δq_2) $\neq$ sign(Δq_1)) (see Table 2.3 for details). The theoretical probability of such case was found to be 0.75 (see Chapter 2.2). In this research, we will use the experimental data of joint error measurements of 2-R robotic manipulator RM 10-01 (SCARA type robot produced by Bulgarian robot manufacturing plant) carried out by Dimov, Dobrinov and Boiadjiev (Dimov, Dobrinov and Boiadjiev, 1997), when the robot positioned into the commanded points, to analyze the efficiency of method of joint error maximum mutual compensation in practice.

Thirty experiments of repeated positioning in each of five points $\mathbf{P}_1$, $\mathbf{P}_2$, $\mathbf{P}_3$, $\mathbf{P}_4$ and $\mathbf{P}_5$ (see Fig. 2.3) of SCARA type robotic manipulator RM 10-01 (see Fig. 2.4 for a simplified kinematic scheme of RM 10-01) with DC (direct current) motors in joint drives (unfortunately, we did not have an experimental setup for industrial robot with stepping motors as joint drives) have been carried out by the authors (Dimov, Dobrinov and Boiadjiev, 1997). The details of the experimental scheme are presented in Chapter 2.1. The industrial robot RM 10-01 has two rotational joints (see Fig. 2.4) and one translational joint, which was not used in the experiments (Dimov, Dobrinov and Boiadjiev, 1997). The translational joint is a pneumatic one with two fixed positions – up and down. The method of control was "point to point" (Dimov, Dobrinov and Boiadjiev, 1997).

Our aim in analyzing these experiments is to verify the theory that the probability of end-effector positioning accuracy improvement, if method of joint error maximum mutual compensation is used for 2-R robotic manipulator with optimal second joint coordinate $q_{2opt} = 180°$, is not less than 0.75 when joint errors are unknown. To verify this, the values of joint errors Δq_1 and Δq_2, measured in the experiment (Dimov, Dobrinov and Boiadjiev, 1997) according to ISO 9283 (1998) during positioning in five points (see Chapter 2.1 for details), are presented in Table 2.4 (Dimov, Dobrinov and Boiadjiev, 1997).

We made the analysis of joint error values in Table 2.4 based on the comparison of joint error Δq_1 and Δq_2 values and signs with the combinations in Table 2.3 for joint error maximum mutual compensation when $q_{2opt} = 180°$. To analyze the presented experimental results in points $\mathbf{P}_1$, $\mathbf{P}_2$, $\mathbf{P}_3$, $\mathbf{P}_4$ and $\mathbf{P}_5$, Tables 2.5, 2.6, 2.7, 2.8 and 2.9 are provided where the combinations of joint errors (values and signs) are ordered according to eight combinations of signs and values from Table 2.3. As a result, based on combinations of joint error Δq_1 and Δq_2 signs and values presented in Tables 2.5, 2.6, 2.7, 2.8 and 2.9 for points $\mathbf{P}_1$, $\mathbf{P}_2$, $\mathbf{P}_3$, $\mathbf{P}_4$ and $\mathbf{P}_5$, one can see that 0 % of obtained joint error value and sign combinations for Δq_1 and Δq_2 would provide the improvement of end-effector positioning accuracy in $\mathbf{P}_1$, 90 % of combinations would provide the improvement of end-effector positioning accuracy in $\mathbf{P}_2$, 83.33 % of combinations would provide the improvement of end-effector positioning accuracy in $\mathbf{P}_3$, 73.3 % of combinations would provide the improvement of end-effector positioning accuracy in $\mathbf{P}_4$ and 100 % of combinations would provide the improvement of end-effector positioning accuracy in $\mathbf{P}_5$ if the developed method of joint error maximum mutual compensation would be used and joint error values would not change significantly in the new working points after changing end-effector location in the direction of $q_{2opt} = 180°$, as explained in Chapter 2.2. The graphs in Fig. 2.12 show the relations of number of positive combinations of Δq_1 and Δq_2 (combinations 1, 2, 3, 4, 5 and 8, see Tables 2.3, 2.5, 2.6, 2.7, 2.8 and 2.9) that would provide end-effector positioning accuracy improvement for $q_{2opt} =$

180° and the number of negative combinations of Δq_1 and Δq_2 (combinations 6, 7, see Tables 2.3, 2.5, 2.6, 2.7, 2.8 and 2.9) that would deteriorate end-effector positioning accuracy.

Table 2.4

Experimentally obtained joint error values in points $\mathbf{P_1}$, $\mathbf{P_2}$, $\mathbf{P_3}$, $\mathbf{P_4}$ and $\mathbf{P_5}$
(last row shows average values of Δq_1 and Δq_2)

	$\mathbf{P_1}$		$\mathbf{P_2}$		$\mathbf{P_3}$		$\mathbf{P_4}$		$\mathbf{P_5}$	
	Δq_1 [rad] x 10^{-6}	Δq_2 [rad] x 10^{-6}	Δq_1 [rad] x 10^{-6}	Δq_2 [rad] x 10^{-6}	Δq_1 [rad] x 10^{-6}	Δq_2 [rad] x 10^{-6}	Δq_1 [rad] x 10^{-6}	Δq_2 [rad] x 10^{-6}	Δq_1 [rad] x 10^{-6}	Δq_2 [rad] x 10^{-6}
1	11.32	-35.74	10.79	2.36	-1.25	5.26	2.75	1.53	-4.47	3.22
2	17.28	-36.08	8.38	0.16	-11.08	7.91	-4.92	7.05	-9.68	1.92
3	19.8	-36.11	6.63	7.6	1.61	-0.25	2.65	3.58	-8.8	3.61
4	12.71	-36.15	3.68	13.94	-3.26	3.21	-3.09	2.57	-2.74	2.17
5	22.79	-36.27	10.12	-7.27	-12	11.18	6.06	3.79	-7.95	0.88
6	18.67	-36.48	12.53	-5.07	-13.01	10.16	-7.89	8.26	-16.63	-2.74
7	26.43	-35.92	17.9	-9.21	4.54	-1.48	-6.75	1.53	-16.65	-7.16
8	23.92	-35.89	12.53	-5.07	3.62	1.8	-6.75	1.53	-19.22	-3.39
9	25.3	-36.3	1.94	21.38	8.49	-1.67	-3.86	11.08	-15.76	-1.05
10	25.26	-34.85	5.42	6.5	0.69	3.01	-3.65	6.78	-19.21	1.02
11	30.33	-36.36	8.38	0.16	1.7	4.04	2.72	4.35	-23.54	1.42
12	33.78	-36.66	4.21	5.4	3.62	1.8	5.65	5.18	-21.81	0.37
13	28.28	-36.46	18.44	-17.75	10.43	-3.92	7.93	8.65	-30.48	1.16
14	28.28	-36.46	6.63	7.6	9.5	-0.65	7.24	0.9	-33.09	0.52
15	23.92	-35.89	20.18	-25.19	2.69	5.06	5.07	5.39	-22.69	-1.30
16	31	-35.84	10.79	2.36	-3.18	7.51	7.89	8.26	-25.29	-1.95
17	23.45	-35.75	5.42	6.5	1.61	-0.25	6.75	1.53	-14.03	-2.09
18	26.89	-36.06	14.27	-12.51	10.43	-3.92	5.65	5.18	-16.61	1.67
19	28.28	-36.46	15.48	-11.41	8.49	-1.67	6.79	1.91	-13.14	4.00
20	33.32	-36.52	14.95	-2.87	11.43	-2.9	4.96	7.43	-13.15	-0.40
21	31.26	-36.63	6.63	7.6	-7.13	7.71	-1.58	17.62	-7.95	0.88
22	25.3	-36.3	3.68	13.94	-7.13	7.71	-0.31	12.36	-14.03	-2.09
23	25.3	-36.3	14.27	-12.51	7.57	1.6	-2.61	14.8	-27.02	-0.91
24	28.74	-36.6	8.38	0.16	-3.18	7.51	3.89	8.26	-31.36	-0.52
25	23.71	-36.54	13.74	-3.97	11.43	-2.9	3.86	11.08	-23.54	1.42
26	20.73	-36.37	9.59	1.26	9.5	-0.65	9	7.81	-28.75	0.12
27	23.71	-36.54	8.38	0.16	1.7	4.04	2.07	5.39	-24.42	-0.26
28	28.28	-36.46	17.22	-18.85	4.63	2.81	4.96	7.43	-31.36	-0.52
29	28.74	-36.6	10.76	-1.27	9.5	-0.65	-1.55	7.23	-26.15	0.77
30	26.22	-36.57	21.39	-24.09	4.63	2.81	5.65	5.18	-27.02	-0.91
Σ/ 30	**25.1**	**-36.23**	**10.76**	**-1.99**	**2.29**	**2.11**	**1.95**	**6.45**	**-19.22**	**-0.02**

Table 2.5

Distribution of joint error values and signs for point $\mathbf{P}_1$ based on Tables 2.3 and 2.4 (*improved* in Definition column means that end-effector positioning accuracy would improve for such combination of joint error signs and values (with $q_{2opt} = 180°$) if joint error maximum mutual compensation would be applied)

№	Δq_1 sign	Δq_2 sign	Relation	q_{2opt}, [rad]	Definition	Number of occurrences
1	+	+	$\lvert\Delta q_2\rvert \leq \lvert\Delta q_1\rvert$ & $\mathrm{sign}\Delta q_2 = \mathrm{sign}\Delta q_1$	π	improved	0
2	+	-	$\lvert\Delta q_2\rvert \leq \lvert\Delta q_1\rvert$ & $\mathrm{sign}\Delta q_2 \neq \mathrm{sign}\Delta q_1$	π	improved	0
3	-	+	$\lvert\Delta q_2\rvert \leq \lvert\Delta q_1\rvert$ & $\mathrm{sign}\Delta q_2 \neq \mathrm{sign}\Delta q_1$	π	improved	0
4	-	-	$\lvert\Delta q_2\rvert \leq \lvert\Delta q_1\rvert$ & $\mathrm{sign}\Delta q_2 = \mathrm{sign}\Delta q_1$	π	improved	0
5	+	+	$\lvert\Delta q_2\rvert \geq \lvert\Delta q_1\rvert$ & $\mathrm{sign}\Delta q_2 = \mathrm{sign}\Delta q_1$	π	improved	0
6	+	-	$\lvert\Delta q_2\rvert > \lvert\Delta q_1\rvert$ & $\mathrm{sign}\Delta q_2 \neq \mathrm{sign}\Delta q_1$	π	**deteriorated**	30
7	-	+	$\lvert\Delta q_2\rvert > \lvert\Delta q_1\rvert$ & $\mathrm{sign}\Delta q_2 \neq \mathrm{sign}\Delta q_1$	π	**deteriorated**	0
8	-	-	$\lvert\Delta q_2\rvert \geq \lvert\Delta q_1\rvert$ & $\mathrm{sign}\Delta q_2 = \mathrm{sign}\Delta q_1$	π	improved	0
Summary: Number of cases when end-effector positioning accuracy would improve: 0 Number of cases when end-effector positioning accuracy would deteriorate: 30 Percentage of cases when end-effector positioning accuracy would improve: 0 Percentage of cases when end-effector positioning accuracy would deteriorate: 100						

Table 2.6

Distribution of joint error values and signs for point $\mathbf{P}_2$ based on Tables 2.3 and 2.4 (*improved* in Definition column means that end-effector positioning accuracy would improve for such combination of joint error signs and values (with $q_{2opt} = 180°$) if joint error maximum mutual compensation would be applied)

№	Δq_1 sign	Δq_2 sign	Relation	q_{2opt}, [rad]	Definition	Number of occurrences
1	+	+	$\lvert\Delta q_2\rvert \le \lvert\Delta q_1\rvert$ & signΔq_2 = signΔq_1	π	improved	7
2	+	-	$\lvert\Delta q_2\rvert \le \lvert\Delta q_1\rvert$ & sign$\Delta q_2 \neq$ signΔq_1	π	improved	11
3	-	+	$\lvert\Delta q_2\rvert \le \lvert\Delta q_1\rvert$ & sign$\Delta q_2 \neq$ signΔq_1	π	improved	0
4	-	-	$\lvert\Delta q_2\rvert \le \lvert\Delta q_1\rvert$ & signΔq_2 = signΔq_1	π	improved	0
5	+	+	$\lvert\Delta q_2\rvert \ge \lvert\Delta q_1\rvert$ & signΔq_2 = signΔq_1	π	improved	9
6	+	-	$\lvert\Delta q_2\rvert > \lvert\Delta q_1\rvert$ & sign$\Delta q_2 \neq$ signΔq_1	π	**deteriorated**	3
7	-	+	$\lvert\Delta q_2\rvert > \lvert\Delta q_1\rvert$ & sign$\Delta q_2 \neq$ signΔq_1	π	**deteriorated**	0
8	-	-	$\lvert\Delta q_2\rvert \ge \lvert\Delta q_1\rvert$ & signΔq_2 = signΔq_1	π	improved	0
Summary: Number of cases when end-effector positioning accuracy would improve: 27 Number of cases when end-effector positioning accuracy would deteriorate: 3 Percentage of cases when end-effector positioning accuracy would improve: 90 Percentage of cases when end-effector positioning accuracy would deteriorate: 10						

Table 2.7

Distribution of joint error values and signs for point $\mathbf{P}_3$ based on Tables 2.3 and 2.4 (*improved* in Definition column means that end-effector positioning accuracy would improve for such combination of joint error signs and values (with $q_{2opt} = 180°$) if joint error maximum mutual compensation would be applied)

№	Δq_1 sign	Δq_2 sign	Relation	q_{2opt}, [rad]	Definition	Number of occurrences
1	+	+	$\lvert\Delta q_2\rvert \leq \lvert\Delta q_1\rvert$ & $\mathrm{sign}\Delta q_2 = \mathrm{sign}\Delta q_1$	π	improved	5
2	+	-	$\lvert\Delta q_2\rvert \leq \lvert\Delta q_1\rvert$ & $\mathrm{sign}\Delta q_2 \neq \mathrm{sign}\Delta q_1$	π	improved	12
3	-	+	$\lvert\Delta q_2\rvert \leq \lvert\Delta q_1\rvert$ & $\mathrm{sign}\Delta q_2 \neq \mathrm{sign}\Delta q_1$	π	improved	4
4	-	-	$\lvert\Delta q_2\rvert \leq \lvert\Delta q_1\rvert$ & $\mathrm{sign}\Delta q_2 = \mathrm{sign}\Delta q_1$	π	improved	0
5	+	+	$\lvert\Delta q_2\rvert \geq \lvert\Delta q_1\rvert$ & $\mathrm{sign}\Delta q_2 = \mathrm{sign}\Delta q_1$	π	improved	4
6	+	-	$\lvert\Delta q_2\rvert > \lvert\Delta q_1\rvert$ & $\mathrm{sign}\Delta q_2 \neq \mathrm{sign}\Delta q_1$	π	**deteriorated**	0
7	-	+	$\lvert\Delta q_2\rvert > \lvert\Delta q_1\rvert$ & $\mathrm{sign}\Delta q_2 \neq \mathrm{sign}\Delta q_1$	π	**deteriorated**	5
8	-	-	$\lvert\Delta q_2\rvert \geq \lvert\Delta q_1\rvert$ & $\mathrm{sign}\Delta q_2 = \mathrm{sign}\Delta q_1$	π	improved	0

Summary:

Number of cases when end-effector positioning accuracy would improve: 25

Number of cases when end-effector positioning accuracy would deteriorate: 5

Percentage of cases when end-effector positioning accuracy would improve: 83.33

Percentage of cases when end-effector positioning accuracy would deteriorate: 16.67

Table 2.8

Distribution of joint error values and signs for point $\mathbf{P}_4$ based on Tables 2.3 and 2.4 (*improved* in Definition column means that end-effector positioning accuracy would improve for such combination of joint error signs and values (with $q_{2opt} = 180°$) if joint error maximum mutual compensation would be applied)

№	Δq_1 sign	Δq_2 sign	Relation	q_{2opt}, [rad]	Definition	Number of occurrences
1	+	+	$\lvert\Delta q_2\rvert \leq \lvert\Delta q_1\rvert$ & $\text{sign}\Delta q_2 = \text{sign}\Delta q_1$	π	improved	10
2	+	-	$\lvert\Delta q_2\rvert \leq \lvert\Delta q_1\rvert$ & $\text{sign}\Delta q_2 \neq \text{sign}\Delta q_1$	π	improved	0
3	-	+	$\lvert\Delta q_2\rvert \leq \lvert\Delta q_1\rvert$ & $\text{sign}\Delta q_2 \neq \text{sign}\Delta q_1$	π	improved	3
4	-	-	$\lvert\Delta q_2\rvert \leq \lvert\Delta q_1\rvert$ & $\text{sign}\Delta q_2 = \text{sign}\Delta q_1$	π	improved	0
5	+	+	$\lvert\Delta q_2\rvert \geq \lvert\Delta q_1\rvert$ & $\text{sign}\Delta q_2 = \text{sign}\Delta q_1$	π	improved	9
6	+	-	$\lvert\Delta q_2\rvert > \lvert\Delta q_1\rvert$ & $\text{sign}\Delta q_2 \neq \text{sign}\Delta q_1$	π	**deteriorated**	0
7	-	+	$\lvert\Delta q_2\rvert > \lvert\Delta q_1\rvert$ & $\text{sign}\Delta q_2 \neq \text{sign}\Delta q_1$	π	**deteriorated**	8
8	-	-	$\lvert\Delta q_2\rvert \geq \lvert\Delta q_1\rvert$ & $\text{sign}\Delta q_2 = \text{sign}\Delta q_1$	π	improved	0

Summary:

Number of cases when end-effector positioning accuracy would improve: 22

Number of cases when end-effector positioning accuracy would deteriorate: 8

Percentage of cases when end-effector positioning accuracy would improve: 73.33

Percentage of cases when end-effector positioning accuracy would deteriorate: 26.67

Table 2.9

Distribution of joint error values and signs for point $\mathbf{P}_5$ based on Tables 2.3 and 2.4 (*improved* in Definition column means that end-effector positioning accuracy would improve for such combination of joint error signs and values (with $q_{2opt} = 180°$) if joint error maximum mutual compensation would be applied)

№	Δq_1 sign	Δq_2 sign	Relation	q_{2opt}, [rad]	Definition	Number of occurrences
1	+	+	$\lvert\Delta q_2\rvert \le \lvert\Delta q_1\rvert$ & $\text{sign}\Delta q_2 = \text{sign}\Delta q_1$	π	improved	0
2	+	-	$\lvert\Delta q_2\rvert \le \lvert\Delta q_1\rvert$ & $\text{sign}\Delta q_2 \ne \text{sign}\Delta q_1$	π	improved	0
3	-	+	$\lvert\Delta q_2\rvert \le \lvert\Delta q_1\rvert$ & $\text{sign}\Delta q_2 \ne \text{sign}\Delta q_1$	π	improved	16
4	-	-	$\lvert\Delta q_2\rvert \le \lvert\Delta q_1\rvert$ & $\text{sign}\Delta q_2 = \text{sign}\Delta q_1$	π	improved	14
5	+	+	$\lvert\Delta q_2\rvert \ge \lvert\Delta q_1\rvert$ & $\text{sign}\Delta q_2 = \text{sign}\Delta q_1$	π	improved	0
6	+	-	$\lvert\Delta q_2\rvert > \lvert\Delta q_1\rvert$ & $\text{sign}\Delta q_2 \ne \text{sign}\Delta q_1$	π	**deteriorated**	0
7	-	+	$\lvert\Delta q_2\rvert > \lvert\Delta q_1\rvert$ & $\text{sign}\Delta q_2 \ne \text{sign}\Delta q_1$	π	**deteriorated**	0
8	-	-	$\lvert\Delta q_2\rvert \ge \lvert\Delta q_1\rvert$ & $\text{sign}\Delta q_2 = \text{sign}\Delta q_1$	π	improved	0

Summary:

Number of cases when end-effector positioning accuracy would improve: 30

Number of cases when end-effector positioning accuracy would deteriorate: 0

Percentage of cases when end-effector positioning accuracy would improve: 100

Percentage of cases when end-effector positioning accuracy would deteriorate: 0

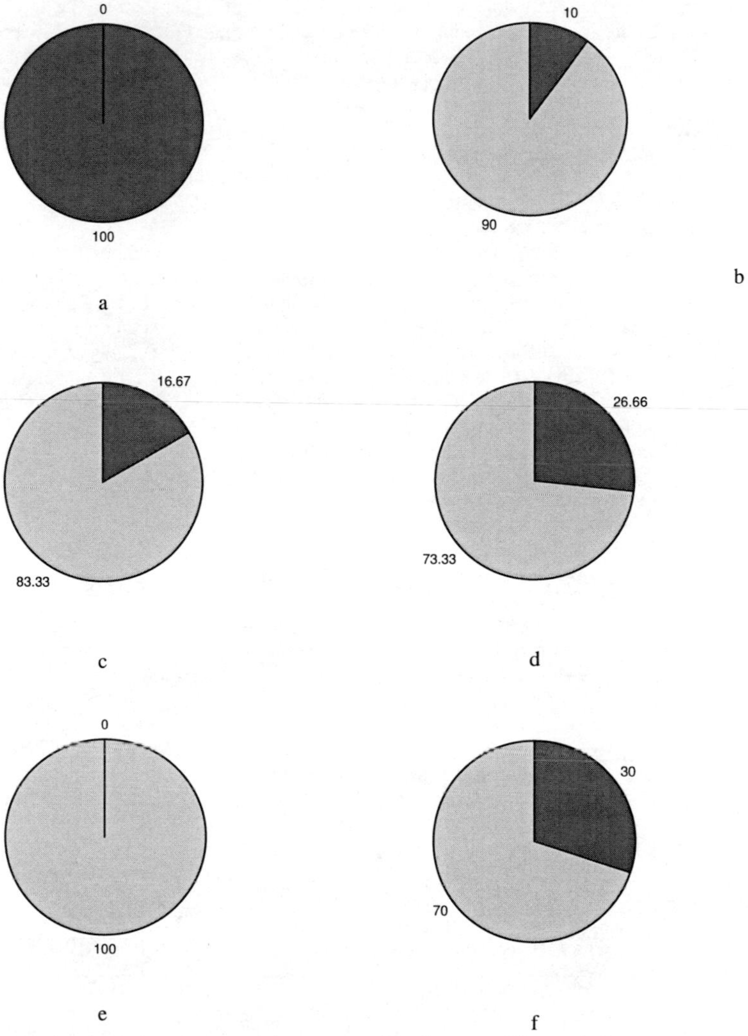

Fig. 2.12. Graphs of relations between joint error combinations providing end-effector positioning accuracy improvement (grey color) and end-effector positioning accuracy deterioration (black color) if joint error maximum mutual compensation would be used with $q_{2opt} = 180°$: a - in point $\mathbf{P}_1$; b - in point $\mathbf{P}_2$; c - in point $\mathbf{P}_3$; d - in point $\mathbf{P}_4$, e - in point $\mathbf{P}_5$ and f – average value for all points $\mathbf{P}_1$, $\mathbf{P}_2$, $\mathbf{P}_3$, $\mathbf{P}_4$ and $\mathbf{P}_5$

As one can see from Fig. 2.12, the results of our analysis prove the theory, namely, the undesirable combinations when joint error $|\Delta q_2|$ is not higher than $|\Delta q_1|$ and joint error signs are not equal ($\text{sign}(\Delta q_2) \neq \text{sign}(\Delta q_1)$) were found in points **P**$_2$, **P**$_3$, **P**$_4$ and **P**$_5$, accordingly in 10 %, 16.67 %, 26.67 % and 0 % of cases. Many of these values are even much lower than the theoretical value of 25 % (probability of 0.25) for the normal Gaussian distribution of joint errors. However, the negative effect was found in point **P**$_1$ where end-effector positioning accuracy would deteriorate in 100 % of cases. The average percentage of end-effector positioning accuracy deterioration for all five points is (100 % + 10 % + 16.67 % + 26.67 % + 0 %) / 5 = 30 %. It is a bit higher than a theoretical value of 25 % (probability of 0.25) for normal Gaussian distribution of joint errors, however, it can still be considered to be acceptable.

As a result, the application of joint error maximum mutual compensation in the given experimental scheme, if joint error values would be unknown, would bring a positive effect (improved end-effector positioning accuracy) in points **P**$_2$, **P**$_3$, **P**$_4$ and **P**$_5$ for q_{2opt} = 180° and the negative effect (deteriorated end-effector positioning accuracy) only in point **P**$_1$ (see Fig. 2.12 for detailed values). Let us also calculate end-effector positioning accuracy improvement in points **P**$_2$, **P**$_3$, **P**$_4$, **P**$_5$ and end-effector positioning accuracy deterioration in point **P**$_1$ that can be reached for robotic manipulator RM 10-01 (see Chapter 2.1 for details of RM 10-01) if q_{2opt} = 180° and assuming that the allowed distance for end-effector position change may be not more than 10 mm (a_{dist} = 10 mm) as it was done in Chapter 2.2. The results of the calculation are presented in Table 2.10. To calculate end-effector positioning accuracy improvement that was obtained for robot RM 10-01 using joint error maximum mutual compensation, one has to do the following:

1. Calculate initial end-effector positioning accuracy ΔL using (2.4) with available l_1, l_2, Δq_1, Δq_2 (see Table 2.4) and initial q_1 and q_2 (ΔL is known for all points from (ISO 9283, 1998), see Table 2.2);
2. Calculate rotation values δq_1 and δq_2 using (2.6) and (2.7);
3. Calculate new q_1 and q_2 by adding δq_1 and δq_2 to initial values q_1 and q_2;
4. Calculate new end-effector positioning accuracy ΔL using (2.4) with new q_1 and q_2.

Table 2.10

Calculation of end-effector positioning accuracy improvement or deterioration using joint error maximum mutual compensation in points **P**$_1$, **P**$_2$, **P**$_3$, **P**$_4$ and **P**$_5$ for industrial robot RM 10-01 (l_1 = 0.4 m and l_2 = 0.25 m, q_{2opt} = 180°, a_{dist} = 10 mm)

Point	Initial q_1 [rad]	Initial q_2 [rad]	Initial ΔL [μm]	δq_1 [rad]	δq_2 [rad]	Final q_1 [rad]	Final q_2 [rad]	Final ΔL [μm]	Improvement / Deterioration [%]
P$_1$	1.17	1.79	10.9	-0.01	0.04	1.16	1.83	11.1	-1.8
P$_2$	1.06	0.9	5.9	0.31	0.07	1.37	0.97	5.83	1.4
P$_3$	1.77	0.79	1.87	-0.01	0.08	1.76	0.87	1.81	3.2
P$_4$	2.11	2.236	1.73	~0	0.05	2.11	2.286	1.69	2.4
P$_5$	0.34	2.215	6.15	~0	0.05	0.34	2.265	5.9	4.1

Based on our analysis of experimental results (Dimov, Dobrinov and Boiadjiev, 1997), we can formulate a conclusion that method of joint error maximum mutual compensation will provide end-effector positioning accuracy improvement for industrial robot RM 10-01 in most cases. If joint error values are not known then with the probability of about 0.75 the use of joint error maximum mutual compensation for 2-R robotic manipulators will improve end-effector positioning accuracy. Most likely, the same results can be predicted for other similar industrial robots.

If one could change the position of robot end-effector within the whole robot workspace and joint errors would remain unchanged (in practice, such case is not common (Kieffer, Cahill and James, 1997) because joint errors may significantly change with the change of the end-effector position in the robot workspace and q_{2opt} = 180° is on the border of the robot workspace), then we could directly use q_{2opt} = 180° and see the theoretical maximum of joint error maximum mutual compensation for industrial robot RM 10-01. Table 2.11 contains the results of calculating theoretical maximum of end-effector positioning improvement or deterioration using joint error maximum mutual compensation for end-effector positioning of industrial robot RM 10-01 in points $\mathbf{P}_1$, $\mathbf{P}_2$, $\mathbf{P}_3$, $\mathbf{P}_4$ and $\mathbf{P}_5$. To calculate theoretical maximum of end-effector positioning accuracy improvement that was obtained for robot RM 10-01 using joint error maximum mutual compensation, one has to do the following:

1. Calculate initial end-effector positioning accuracy ΔL using (2.4) with available l_1, l_2, Δq_1, Δq_2 (see Table 2.4) and initial q_1 and q_2 (ΔL is known for all points from (Dimov, Dobrinov and Boiadjiev, 1997), see Table 2.2);
2. Change coordinate q_2 so that it is equal to q_{2opt} = 180°;
3. Calculate new end-effector positioning accuracy ΔL using (2.4) with new q_2 = 180°.

Table 2.11

Calculation of theoretical maximum of end-effector positioning accuracy improvement or deterioration using joint error maximum mutual compensation in points $\mathbf{P}_1$, $\mathbf{P}_2$, $\mathbf{P}_3$, $\mathbf{P}_4$ and $\mathbf{P}_5$ for industrial robot RM 10-01 (l_1 = 0.4 m and l_2 = 0.25 m, q_{2opt} = 180°)

Point	Initial q_1 [rad]	Initial q_2 [rad]	Initial ΔL [μm]	Final q_2 [rad]	Final ΔL [μm]	Improvement / Deterioration [%]
$\mathbf{P}_1$	1.17	1.79	10.9	π	12.8	-17.4
$\mathbf{P}_2$	1.06	0.9	5.9	π	2.11	64.3
$\mathbf{P}_3$	1.77	0.79	1.87	π	0.18	90.4
$\mathbf{P}_4$	2.11	2.236	1.73	π	1.32	23.7
$\mathbf{P}_5$	0.34	2.215	6.15	π	2.87	53.4

As one can see from Table 2.11, the theoretical maximum of end-effector positioning accuracy improvement using joint error maximum mutual compensation in points $\mathbf{P}_2$, $\mathbf{P}_3$, $\mathbf{P}_4$ and $\mathbf{P}_5$ is quite high (for example, it could be 90.4 % for point $\mathbf{P}_3$). However, end-effector positioning accuracy would deteriorate in point $\mathbf{P}_1$ by 17.4 %.

2.4. GLOBAL APPROACH TO ROBOT JOINT ERROR MAXIMUM MUTUAL COMPENSATION

The joint error maximum mutual compensation (see Chapter 2.2) suggests that a correction of a relative position of manipulator joints is done by changing the coordinate of the second joint in 2-R robotic manipulator in the direction of the joint coordinates under which better joint error mutual compensation takes place, for example, $q_{2opt} = 180°$ (see Chapter 2.2 for details). This can be used not only when joint errors are known but also when joint errors are partly known or not known at all. In the latter two cases, it is supposed that the probability of the case when $|\Delta q_2| > |\Delta q_1|$ and Δq_1 with Δq_2 have opposite signs is at least not higher than 0.25 (see Chapter 2.3). The following conditions are important for joint error maximum mutual compensation:

- Joint error distribution should be Gaussian (normal), because one can use means (average values) for operating with joint errors in this case.
- Joint errors should not change significantly with the slight change of the joint trajectory (precise values are largely dependent on the given robotic manipulator).

The joint error maximum mutual compensation has been presented as applied to a simple 2-R robotic manipulator (see Chapter 2.2). The scheme demonstrating the application of the joint error maximum mutual compensation for 6-DOF robotic manipulator (this is an experimental robotic manipulator that is composed of the SCARA arms) is shown in Fig. 2.13.

The callouts in Fig. 2.13 show that 6-DOF type robotic manipulator has two times two rotational joints in different planes (one of them is in z_0x_0 plane and the other is in z_0y_0). To apply joint error maximum mutual compensation for such manipulator we just have to consider that the robotic manipulator includes 2-R chains (see callouts in Fig. 2.13) in its kinematic scheme and apply step-by-step joint error maximum mutual compensation to each of these 2-R chains separately (as it has been done for 2-R SCARA type robotic manipulator which is a 2-R chain in this case, see Chapter 2.2 for details). 2-R chains in 6-DOF type robotic manipulator (see Fig. 2.13) are formed by links 2, 3 and links 5, 6.

It is important to notice that if joint errors are not known or only partly known in complex robotic manipulators that include 2-R chains then the probability p_{improv} of end-effector positioning accuracy improvement in all 2-R chains for a given complex robotic manipulator using joint error maximum mutual compensation will decrease with the increase of the number of 2-R chains, because p_{improv} can be expressed as:

$$p_{improv} = p_{improv1}\, p_{improv2} \cdots p_{improvN}\,, \tag{2.8}$$

where $p_{improv1}$, $p_{improv2}$ ... $p_{improvN}$ are the probabilities of end-effector positioning accuracy improvement in each of 2-R chains ($p_{improv1} = p_{improv2} = p_{improvN} = 0.75$; see Chapter 2.2 for details) and N is the number of 2-R chains in the given complex robotic manipulator. The probability $p_{deterior}$ of end-effector positioning accuracy deterioration in all 2-R chains for a given complex robotic manipulator using joint error maximum mutual compensation will decrease with the increase of the number of 2-R chains as well, because $p_{deterior}$ can be expressed as:

$$p_{deterior} = p_{deterior1}\, p_{deterior2} \cdots p_{deteriorN}\,, \tag{2.9}$$

where $p_{deterior1}$, $p_{deterior2}$ ... $p_{deteriorN}$ are the probabilities of end-effector positioning accuracy deterioration in each of 2-R chains ($p_{deterior1} = p_{deterior2} = p_{deteriorN} = 0.25$; see Chapter 2.2 for details) and N is the number of 2-R chains in the given complex robotic manipulator.

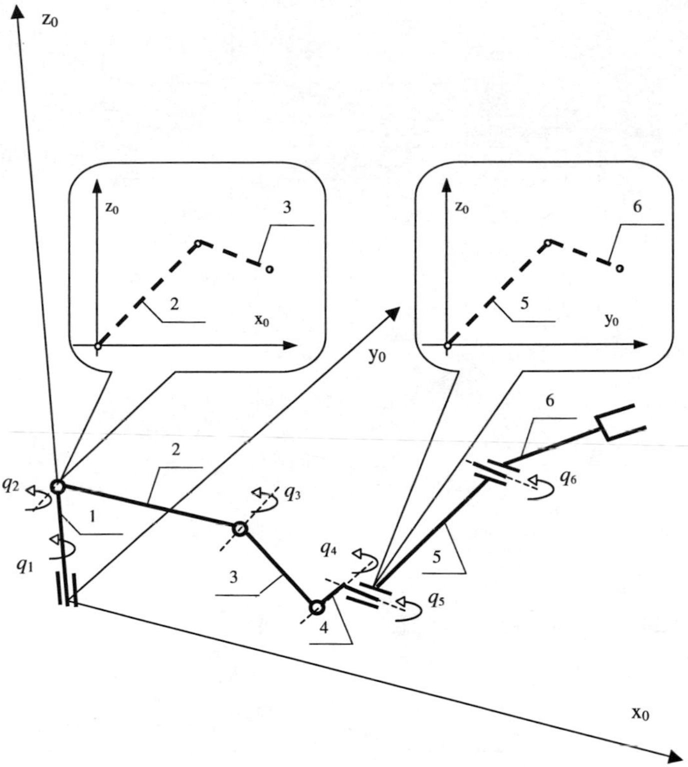

Fig. 2.13. The kinematic scheme of 6 - DOF robotic manipulator and its 2-R chains (shown in callouts) formed by links 2, 3 and links 5, 6

The probability p_{others} of other combinations (this means combinations in which in some 2-R chains the improvement and in some the deterioration will take place) can be expressed as:

$$p_{others} = 1 - (p_{deterior} + p_{improve}), \tag{2.10}$$

Appropriate graphs of dependencies for probabilities p_{others}, $p_{deterior}$ and $p_{improve}$ on the number of 2-R chains N are shown in Fig. 2.14.

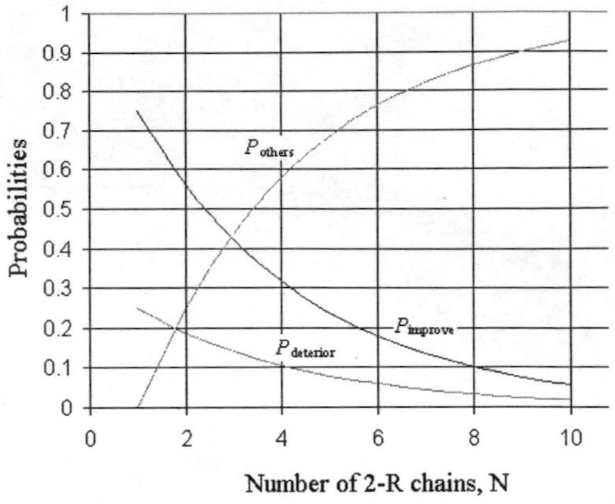

Fig. 2.14. Dependencies for probabilities p_{others}, $p_{deterior}$ and $p_{improve}$ on the number of 2-R chains N

For 6-DOF robotic manipulator (see Fig. 2.13) using (2.8), (2.9) and (2.10), one can calculate that $p_{improv} = p_{improv1}\ p_{improv2} = 0.75 \times 0.75 = 0.5625$, $p_{deterior} = p_{deterior1}\ p_{deterior2} = 0.25 \times 0.25 = 0.0625$ and $p_{others} = 1 - (p_{deterior} + p_{improve}) = 1 - (0.0625 + 0.5625) = 0.375$. Probability p_{others} for 6-DOF robotic manipulator (see Fig. 2.13) is when in one 2-R chain the improvement of end-effector positioning accuracy takes place and in the other 2-R chain the deterioration of end-effector positioning accuracy takes place.

As one can see from graphs in Fig. 2.14, the implementation of joint error maximum mutual compensation is the most effective for 2-R SCARA type robotic manipulators (see Fig. 1.3), because with the increase of the number of 2-R chains the probability p_{improv} decreases significantly and already for robotic manipulator with three 2-R chains it is only 0.42 (lower than 0.5). The probability p_{others} increases significantly and brings additional complication to the overall end-effector positioning accuracy evaluation (improvement or deterioration).

2.5. JOINT ERROR MAXIMUM MUTUAL COMPENSATION IN ROBOTIC MANIPULATORS - WORST CASE ERROR APPROACH

The comparison of results of end-effector positioning accuracy improvement using joint error maximum mutual compensation in points $\mathbf{P}_1$, $\mathbf{P}_2$, $\mathbf{P}_3$, $\mathbf{P}_4$ and $\mathbf{P}_5$ for industrial robot RM 10-01 (see Tables 2.10 and 2.11) shows a potential for further improvement of the proposed local optimization approach (see Chapter 2.2). The best end-effector positioning accuracy improvement achieved in the local optimization approach (with limited use of robot workspace and a_{dist} = 10 mm, q_{2opt} = 180° (see Table 2.10)) was 4.1 % in point $\mathbf{P}_5$. This is a relatively small improvement. The best end-effector positioning accuracy improvement achieved in the global optimization approach with q_2 = 180° (see Table 2.11) was 90.4 % in point $\mathbf{P}_3$. Such improvement could be a very good achievement for use of joint error maximum mutual compensation. However, one should not forget that the global optimization approach in its current form is not possible to be implemented in practice, because the robot configuration with $q_2 = 180°$ is useless in practice and joint error values may change significantly in the new point with $q_2 = 180°$ (see, for example, Table 2.4 to compare how joint error values differ in various points in the robot workspace). We think that the best approach (let us call it worst case error approach) to solve the above-mentioned problem for joint error maximum mutual compensation could be found somewhere in between local and global optimization approaches, as discussed further.

We propose to use the worst case error approach, which is to extend the allowed change a_{dist} of the initial working point $\mathbf{P}_{initial}$, so that the value of a_{dist} could be much higher than previously pre-defined value (*e.g.*, about 10 mm for robot RM 10-01). Unfortunately, in this case we cannot guarantee any more that values of joint errors will not change significantly in the new point $\mathbf{P}_{final}$ located far from the initial point $\mathbf{P}_{initial}$ (for example, see Table 2.4). How joint error values will change are hard to predict because there are various factors that influence on them (*e.g.*, coordinate transformation errors, external influences such as temperature, differences between the dimensions of the articulated structure and those used in the robot control system model, mechanical faults such as clearances, hysteresis, friction, *etc.*). Much more, in Chapter 2.2 we stated that we aimed at using joint error maximum mutual compensation even when joint error values are not known or only partly known. The solution could be to allow such location change a_{dist} of the initial working point $\mathbf{P}_{initial}$ that guarantees that joint error values are within the empirically (or experimentally) found interval:

$$\Delta q_i^{\,l} \leq \Delta q_i \leq \Delta q_i^{\,r}, \tag{2.11}$$

where $\Delta q_i^{\,l}$ is the left limit of the joint error in the i-th joint and $\Delta q_i^{\,r}$ is the right limit of the joint error in the i-th joint. However, we will still not know how joint errors change with the change of the joint coordinates (*i.e.*, we will not know $\Delta q_i(\mathbf{q})$).

The benefits of the worst case error approach (see (2.11)) in comparison to the local optimization approach (see Chapter 2.2) is that we have more freedom in changing the location of the initial working point $\mathbf{P}_{initial}$ and, thus, can move much closer towards $q_{2opt} =$ 180° with the best end-effector positioning accuracy improvement (see Table 2.11). The end-effector positioning accuracy improvement by 20 – 30 % for worst case error approach could be a significant progress in comparison to the local optimization approach with end-effector positioning accuracy improvement of about 4 %.

Let us define the theoretical basis for worst case error approach. Equation (2.4) can be a good starting point to explain in more details key aspects of the worst case error approach. In the local optimization approach we agreed that to find the critical points (maximum and minimum) of (2.4) we would suppose that joint errors were not dependent on the joint coordinates (which is still correct if joint coordinates are not changed significantly (see Chapter 2.2 for details)). If we would not agree with the previous statement then we would have to consider that Δq_1 and Δq_2 are dependent on q_1 and q_2 (*i.e.*, Δq_1 (q_1, q_2) and Δq_2 (q_1, q_2)) and to find the critical points of (2.4) we would have to use two equations formed by taking partial derivatives of ΔL^2 with respect to q_1 and with respect to q_2.

$$\begin{aligned} \frac{\partial(\Delta L^2)}{\partial q_1} &= 0 \\ \frac{\partial(\Delta L^2)}{\partial q_2} &= 0\,. \end{aligned} \tag{2.12}$$

(2.12) is a standard solution to find critical points (Gmurman, 2000), because $\Delta L^2 = f(q_1, q_2)$ is a function of two variables q_1 and q_2. Equations (2.12) could help us to find the minimum of (2.4) provided that Δq_1 (q_1, q_2) and Δq_2 (q_1, q_2) are known. Unfortunately, it is not the case (q_1 (q_1, q_2) and Δq_2 (q_1, q_2) are not known) and, thus, we will have to do the same step for worst case error approach as we did for local optimization in Chapter 2.2, namely, consider that Δq_1 and Δq_2 are not dependent on q_1 and q_2 and use (2.5) to find the same critical points as those presented in Table 2.3. The difference between worst case error approach and local optimization approach comes when we allow in worst case error approach to change the location of the initial working point $\mathbf{P}_{\text{initial}}$ significantly, provided that we can guarantee that joint errors are within the known limits (see (2.11)). If so, we can use limits Δq_i^{l} and Δq_i^{r} as worst case joint errors to perform with them the same calculations as we did in Table 2.10 to find end-effector positioning accuracy improvement using joint error maximum mutual compensation. In fact, the global algorithm (see Algorithm 2.2) for joint error maximum mutual compensation remains unchanged. We will only change the way of searching for joint coordinates (see Step 4 in Algorithm 2.2) under which joint error maximum mutual compensation takes place.

Let us consider our example with industrial robot RM 10-01 (see Chapter 2.2) to explain the usage of the worst case error approach in more details. We propose to use the following conditions to make our explanation more clear:

1. We will not use the limitation a_{dist} and suppose that any reasonable new joint coordinate q_2 in the final point $\mathbf{P}_{\text{final}}$ is satisfactory. Since $q_{2\text{opt}} = 180°$ cannot be used in practice in conventional robotic manipulators, let us use $q_2 = 135°$ as optimal joint coordinate in $\mathbf{P}_{\text{final}}$ ($q_2 = 135°$ was selected empirically because it is relatively close to $q_{2\text{opt}} = 180°$ and still relatively far away from the border of robot workspace). This means that in the given example the distance between $\mathbf{P}_{\text{initial}}$ and $\mathbf{P}_{\text{final}}$ is not important for us. On the other hand, one could, of course, implement the same approach as we used with a_{dist} in Chapter 2.2.
2. Since we do not use a_{dist} in our example, we do not have to change joint coordinate q_1 (*i.e.*, $\delta q_1 = 0$). We can say then that Δq_1 will not change in the new point $\mathbf{P}_{\text{final}}$ (*i.e.*, $\Delta q_1^{\text{l}} = \Delta q_1^{\text{r}} = \Delta q_1$), because the trajectory for the first joint coordinate will not change significantly (Kieffer, Cahill and James, 1997).
3. We will change only second joint coordinate q_2 using δq_2 (see Fig. 2.10), provided that $q_2 < 135°$, otherwise we assume that no improvement of end-effector positioning accuracy using joint error maximum mutual compensation is possible. The required second joint coordinate change δq_2 can be found as:

$$\delta q_2 = 135° - q_2 , \tag{2.13}$$

   where q_2 is the second joint coordinate in the initial point $\mathbf{P}_{\text{initial}}$.
4. To define Δq_2^{l} and Δq_2^{r}, we will use the statistical data from previous experiments (see Table 2.4). In this case, we can set $\Delta q_2^{\text{l}} = -36.23 \times 10^{-6}$ rad and $\Delta q_2^{\text{r}} = 6.45 \times 10^{-6}$ rad, because these were the average critical values observed during positioning in points $\mathbf{P}_1$, $\mathbf{P}_2$, $\mathbf{P}_3$, $\mathbf{P}_4$ and $\mathbf{P}_5$. If no statistical data are available then one has to use empirical values.

With the above mentioned conditions, the results of calculations of end-effector positioning accuracy improvement or deterioration using joint error maximum mutual compensation for industrial robot RM 10-01 in points $\mathbf{P}_1$, $\mathbf{P}_2$, $\mathbf{P}_3$, $\mathbf{P}_4$ and $\mathbf{P}_5$ (see Chapter 2.3) with worst case joint errors $\Delta q_2^{\text{l}} = -36.23 \times 10^{-6}$ rad and $\Delta q_2^{\text{r}} = 6.45 \times 10^{-6}$ rad are presented in Tables 2.12 and

2.13. Table 2.14 contains the average case scenario, namely, when $\Delta q_2 = (\Delta q_2^{l} + \Delta q_2^{r}) / 2$ in the new point $\mathbf{P}_{final}$.

Table 2.12

Calculation of end-effector positioning accuracy improvement or deterioration using joint error maximum mutual compensation in points $\mathbf{P}_1$, $\mathbf{P}_2$, $\mathbf{P}_3$, $\mathbf{P}_4$ and $\mathbf{P}_5$ for industrial robot RM 10-01 ($l_1 = 0.4$ m and $l_2 = 0.25$ m, $q_{2opt} = 135°$) with $\Delta q_2 = \Delta q_2^{l} = -36.23 \times 10^{-6}$ rad in the new point $\mathbf{P}_{final}$

Point	Initial q_1 [rad]	Initial q_2 [rad]	Initial Δq_1 [rad] x 10^{-6}	Initial Δq_2 [rad] x 10^{-6}	Initial ΔL [μm]	δq_2 [rad] x 10^{-6}	Final q_2 [rad]	Final ΔL [μm]	Improvement / Deterioration [%]
$\mathbf{P}_1$	1.17	1.79	25.1	-36.2	10.9	0.57	$\frac{3}{4}\pi$	12.1	-11.0
$\mathbf{P}_2$	1.06	0.9	10.7	-1.9	5.9	1.46	$\frac{3}{4}\pi$	5.0	15.3
$\mathbf{P}_3$	1.77	0.79	2.2	2.1	1.87	1.57	$\frac{3}{4}\pi$	9.1	-386.6
$\mathbf{P}_4$	2.11	2.236	1.9	6.4	1.73	0.12	$\frac{3}{4}\pi$	9.13	-427.7
$\mathbf{P}_5$	0.34	2.215	-19.2	-0.02	6.15	0.15	$\frac{3}{4}\pi$	10.0	-62.6

The analysis of results presented in Tables 2.12 – 2.14 allows making the following conclusions:

- For all points $\mathbf{P}_1$, $\mathbf{P}_2$, $\mathbf{P}_3$, $\mathbf{P}_4$ and $\mathbf{P}_5$, end-effector positioning accuracy improvement has been achieved when $\Delta q_2 = \Delta q_2^{r} = 6.45 \times 10^{-6}$ rad in the new point $\mathbf{P}_{final}$. The best result was 44.1 % end-effector positioning accuracy improvement in point $\mathbf{P}_2$ (see Table 2.13).
- Unfortunately, for all points except $\mathbf{P}_2$, end-effector positioning accuracy deterioration has been observed when $\Delta q_2 = \Delta q_2^{l} = -36.23 \times 10^{-6}$ rad in the new point $\mathbf{P}_{final}$. The worst result was - 427.7 % end-effector positioning accuracy deterioration in point $\mathbf{P}_4$ (see Table 2.12).
- The average case scenario (see Table 2.14) shows a mixed view (*i.e.*, end-effector positioning accuracy improvement in points $\mathbf{P}_1$, $\mathbf{P}_2$ and end-effector positioning accuracy deterioration in points $\mathbf{P}_3$, $\mathbf{P}_4$ and $\mathbf{P}_5$).

One should say that end-effector positioning accuracy deterioration dominates in the average case scenario (see Table 2.14). The obtained results have been predictable, because joint errors Δq_2^{l} and Δq_2^{r} differed largely from the initial ones in a number of points (*e.g.*, in some cases, Δq_2^{l} and Δq_2^{r} were up to 20 times larger than initial Δq_2). Even if we could achieve the ideal theoretical maximum (90.4 % improvement of end-effector positioning accuracy in point $\mathbf{P}_3$ (see Table 2.11)) of joint error maximum mutual compensation, one would not be able to compensate such significant changes of joint error Δq_2 in the new point $\mathbf{P}_{final}$ with q_2 =135°.

Table 2.13

Calculation of end-effector positioning accuracy improvement or deterioration using joint error maximum mutual compensation in points **P**$_1$, **P**$_2$, **P**$_3$, **P**$_4$ and **P**$_5$ for industrial robot RM 10-01 (l_1 = 0.4 m and l_2 = 0.25 m, q_{2opt} =135°) with $\Delta q_2 = \Delta q_2^r = 6.45 \times 10^{-6}$ rad in the new point **P**$_{final}$

Point	Initial q_1 [rad]	Initial q_2 [rad]	Initial Δq_1 [rad] x 10^{-6}	Initial Δq_2 [rad] x 10^{-6}	Initial ΔL [µm]	δq_2 [rad] x 10^{-6}	Final q_2 [rad]	Final ΔL [µm]	Improvement / Deterioration [%]
P$_1$	1.17	1.79	25.1	-36.2	10.9	0.57	$\frac{3}{4}\pi$	7.1	34.8
P$_2$	1.06	0.9	10.7	-1.9	5.9	1.46	$\frac{3}{4}\pi$	3.3	44.1
P$_3$	1.77	0.79	2.2	2.1	1.87	1.57	$\frac{3}{4}\pi$	1.66	11.3
P$_4$	2.11	2.23	1.9	6.4	1.73	0.12	$\frac{3}{4}\pi$	1.64	5.2
P$_5$	0.34	2.21	-19.2	-0.02	6.15	0.15	$\frac{3}{4}\pi$	5.86	4.7

Our example with results presented in Tables 2.12 – 2.14 showed that in order to benefit from joint error maximum mutual compensation with the worst case error approach (*e.g.*, to achieve end-effector positioning accuracy improvement of 20 % and higher in the worst case scenario) we would have to set more strict requirements on $\Delta^l q_2$ and $\Delta^r q_2$ (*e.g.*, something similar to $\Delta q_2^l \leq 0.8\ \Delta q_2$ and $\Delta q_2^r \leq 1.2\ \Delta q_2$, empirically found), despite the fact that it may not be always possible in real applications of industrial robots.

The dependence of end-effector positioning accuracy improvement (or deterioration) δL in the worst case scenario of joint error maximum mutual compensation on the allowed change δq_2 of the second joint coordinate in the direction of q_{2opt} = 180° and known $\delta\Delta q_2^l = \Delta q_2 - \Delta q_2^l$ or $\delta\Delta q_2^r = \Delta q_2^r - \Delta q_2$ (depending on which of them produces the worst case scenario) can be very useful to make a decision if it makes sense to use joint error maximum mutual compensation with worst case error approach for the given industrial robot application or not.

If actual Δq_2^l and Δq_2^r differ largely, as it was the case in the example with industrial robot RM 10-01 (see Tables 2.12 - 2.14), then the above-mentioned dependence $\delta L(\delta q_2, \delta\Delta q_2^l)$ or $\delta L(\delta q_2, \delta\Delta q_2^r)$ would show us that we are unable to achieve the desired end-effector positioning accuracy improvement using joint error maximum mutual compensation with the worst case error approach, because, for example, the required and actual values of joint error limits have the following relation:

$$\Delta q_2^{l(req)} > \Delta q_2^{l(act)} \text{ or } \Delta q_2^{r(req)} < \Delta q_2^{r(act)}, \qquad (2.14)$$

where $\Delta q_2^{\mathrm{l(act)}}$ and $\Delta q_2^{\mathrm{r(act)}}$ are actual left / right joint error limits found from experiments, statistical data or empirical calculations, $\Delta q_2^{\mathrm{l(req)}}$ and $\Delta q_2^{\mathrm{r(req)}}$ are the required left / right joint error limits to achieve the given end-effector positioning accuracy improvement δL. To achieve the desired end-effector positioning accuracy improvement δL using joint error maximum mutual compensation with the worst case error approach, the following relations have to be satisfied:

$$\Delta q_2^{\mathrm{l(req)}} < \Delta q_2^{\mathrm{l(act)}} \text{ or } \Delta q_2^{\mathrm{r(req)}} > \Delta q_2^{\mathrm{r(act)}} . \tag{2.15}$$

Table 2.14

Average case scenario of end-effector positioning accuracy improvement or deterioration using joint error maximum mutual compensation in points $\mathbf{P}_1$, $\mathbf{P}_2$, $\mathbf{P}_3$, $\mathbf{P}_4$ and $\mathbf{P}_5$ for industrial robot RM 10-01 when $\Delta q_2 = (\Delta q_2^{\mathrm{l}} + \Delta q_2^{\mathrm{r}}) / 2$ in $\mathbf{P}_{\mathrm{final}}$

Point	Initial q_1 [rad]	Initial q_2 [rad]	Initial Δq_1 [rad] x 10^{-6}	Initial Δq_2 [rad] x 10^{-6}	Initial ΔL [μm]	δq_2 [rad] x 10^{-6}	Final q_2 [rad]	Final ΔL [μm]	Improvement / Deterioration [%]
$\mathbf{P}_1$	1.17	1.79	25.1	-36.2	10.9	0.57	$\frac{3}{4}\pi$	9.6	11.9
$\mathbf{P}_2$	1.06	0.9	10.7	-1.9	5.9	1.46	$\frac{3}{4}\pi$	4.15	29.7
$\mathbf{P}_3$	1.77	0.79	2.2	2.1	1.87	1.57	$\frac{3}{4}\pi$	5.38	-187.6
$\mathbf{P}_4$	2.11	2.236	1.9	6.4	1.73	0.12	$\frac{3}{4}\pi$	5.38 5	-211.2
$\mathbf{P}_5$	0.34	2.215	-19.2	-0.02	6.15	0.15	$\frac{3}{4}\pi$	7.93	-28.9

Dependences $\delta\Delta q_2^{\mathrm{l}}(\delta L, \delta q_2)$ and $\delta\Delta q_2^{\mathrm{r}}(\delta L, \delta q_2)$ can be useful to find the required left / right joint error limits to achieve the desired end-effector positioning accuracy improvement δL using joint error maximum mutual compensation with the worst case error approach. We will find the dependences $\delta L(\delta q_2, \delta\Delta q_2^{\mathrm{l}})$, $\delta L(\delta q_2, \delta\Delta q_2^{\mathrm{r}})$, $\delta\Delta q_2^{\mathrm{l}}(\delta L, \delta q_2)$ and $\delta\Delta q_2^{\mathrm{r}}(\delta L, \delta q_2)$ using (2.4). First we define δL as:

$$\delta L = \Delta L_{\mathrm{initial}} - \Delta L_{\mathrm{final}} , \tag{2.16}$$

where $\Delta L_{\mathrm{initial}}$ is the end-effector positioning accuracy in the initial point $\mathbf{P}_{\mathrm{initial}}$ and $\Delta L_{\mathrm{final}}$ is the end-effector positioning accuracy in the final point $\mathbf{P}_{\mathrm{final}}$. Taking into account that

$$q_2^{\mathrm{final}} = q_2 + \delta q_2$$

$$\delta\Delta q_2^{\mathrm{l}} = \Delta q_2 - \Delta q_2^{\mathrm{l}}$$

$$\delta\Delta q_2^{\mathrm{r}} = \Delta q_2^{\mathrm{r}} - \Delta q_2 ,$$

where Δq_2 is the joint error in the second joint in the initial point $\mathbf{P}_{\text{initial}}$, one can easily find $\delta L(\delta q_2, \delta\Delta q_2^{l})$ and $\delta L(\delta q_2, \delta\Delta q_2^{r})$ using (2.4) (we assume that Δq_1 remains unchanged between points $\mathbf{P}_{\text{initial}}$ and $\mathbf{P}_{\text{final}}$, because we will not change joint coordinate q_1):

$$\delta L\,(\delta q_2, \delta\Delta q_2^{l}) = \Delta L_{\text{initial}} - \Delta L_{\text{final}}$$

$$= \sqrt{\Delta q_1^{\,2}(l_1^{\,2}+l_2^{\,2}) + 2l_1l_2(\Delta q_1^{\,2}+\Delta q_1\Delta q_2)\cos q_2 + l_2^{\,2}\Delta q_2^{\,2} + 2\Delta q_1\Delta q_2 l_2^{\,2}} -$$

$$\sqrt{\Delta q_1^{\,2}(l_1^{\,2}+l_2^{\,2}) + 2l_1l_2(\Delta q_1^{\,2}+\Delta q_1(\Delta q_2-\delta\Delta q_2^{\,l}))\cos(q_2+\delta q_2) + l_2^{\,2}(\Delta q_2-\delta\Delta q_2^{\,l})^2 + 2\Delta q_1(\Delta q_2-\delta\Delta q_2^{\,l})l_2^{\,2}}. \tag{2.17}$$

$$\delta L\,(\delta q_2, \delta\Delta q_2^{r}) = \Delta L_{\text{initial}} - \Delta L_{\text{final}}$$

$$= \sqrt{\Delta q_1^{\,2}(l_1^{\,2}+l_2^{\,2}) + 2l_1l_2(\Delta q_1^{\,2}+\Delta q_1\Delta q_2)\cos q_2 + l_2^{\,2}\Delta q_2^{\,2} + 2\Delta q_1\Delta q_2 l_2^{\,2}} -$$

$$\sqrt{\Delta q_1^{\,2}(l_1^{\,2}+l_2^{\,2}) + 2l_1l_2(\Delta q_1^{\,2}+\Delta q_1(\Delta q_2+\delta\Delta q_2^{\,r}))\cos(q_2+\delta q_2) + l_2^{\,2}(\Delta q_2+\delta\Delta q_2^{\,r})^2 + 2\Delta q_1(\Delta q_2+\delta\Delta q_2^{\,r})l_2^{\,2}}. \tag{2.18}$$

To find $\delta\Delta q_2^{l}(\delta L, \delta q_2)$ and $\delta\Delta q_2^{r}(\delta L, \delta q_2)$ we will have to do some additional transformations with (2.4). First, we use only the left side of (2.4) in points $\mathbf{P}_{\text{initial}}$ and $\mathbf{P}_{\text{final}}$:

$$\Delta L_{\text{initial}}^{2} - \Delta L_{\text{final}}^{2} = \Delta L_{\text{initial}}^{2} - (\Delta L_{\text{initial}} - \delta L)^2 = \Delta L_{\text{initial}}^{2} - \Delta L_{\text{initial}}^{2} + 2\Delta L_{\text{initial}}\,\delta L - \delta L^2$$

$$= 2\Delta L_{\text{initial}}\,\delta L - \delta L^2\,. \tag{2.19}$$

Second, let us do additional transformations with $\Delta L_{\text{final}}^{2}(\delta\Delta q_2^{l})$ and $\Delta L_{\text{final}}^{2}(\delta\Delta q_2^{r})$:

$$\Delta L_{\text{final}}^{2}(\delta\Delta q_2^{l}) = \Delta q_1^{\,2}(l_1^{\,2}+l_2^{\,2}) + 2l_1l_2(\Delta q_1^{\,2}+\Delta q_1(\Delta q_2-\delta\Delta q_2^{\,l}))\cos(q_2+\delta q_2) +$$

$$l_2^{\,2}(\Delta q_2-\delta\Delta q_2^{\,l})^2 + 2\Delta q_1(\Delta q_2-\delta\Delta q_2^{\,l})l_2^{\,2}$$

$$= \Delta q_1^{\,2}(l_1^{\,2}+l_2^{\,2}) + 2l_1l_2\Delta q_1^{\,2}\cos(q_2+\delta q_2) + 2l_1l_2\Delta q_1(\Delta q_2-\delta\Delta q_2^{\,l})\cos(q_2+\delta q_2) +$$

$$l_2^{\,2}\Delta q_2^{\,2} - 2l_2^{\,2}\Delta q_2\delta\Delta q_2^{\,l} + l_2^{\,2}\delta\Delta q_2^{\,l2} + 2l_2^{\,2}\Delta q_1\Delta q_2 - 2l_2^{\,2}\Delta q_1\delta\Delta q_2^{\,l}$$

$$= \Delta q_1^{\,2}(l_1^{\,2}+l_2^{\,2}) + 2l_1l_2\Delta q_1^{\,2}\cos(q_2+\delta q_2) + 2l_1l_2\Delta q_1\Delta q_2\cos(q_2+\delta q_2) -$$

$$2l_1l_2\Delta q_1\delta\Delta q_2^{\,l}\cos(q_2+\delta q_2) + l_2^{\,2}\Delta q_2^{\,2} - 2l_2^{\,2}\Delta q_2\delta\Delta q_2^{\,l} + l_2^{\,2}\delta\Delta q_2^{\,l2} +$$

$$2l_2^2\Delta q_1\Delta q_2 - 2l_2^2\Delta q_1\delta\Delta q_2^{l}. \qquad (2.20)$$

$$\Delta L_{final}^2(\delta\Delta q_2^{r}) = \Delta q_1^2(l_1^2+l_2^2)+2l_1l_2(\Delta q_1^2+\Delta q_1(\Delta q_2+\delta\Delta q_2^{r}))\cos(q_2+\delta q_2)+$$

$$l_2^2(\Delta q_2+\delta\Delta q_2^{r})^2+2\Delta q_1(\Delta q_2+\delta\Delta q_2^{r})l_2^2$$

$$=\Delta q_1^2(l_1^2+l_2^2)+2l_1l_2\Delta q_1^2\cos(q_2+\delta q_2)+2l_1l_2\Delta q_1(\Delta q_2+\delta\Delta q_2^{r})\cos(q_2+\delta q_2)+$$

$$l_2^2\Delta q_2^2+2l_2^2\Delta q_2\delta\Delta q_2^{r}+l_2^2\delta\Delta q_2^{r2}+2l_2^2\Delta q_1\Delta q_2+2l_2^2\Delta q_1\delta\Delta q_2^{r}$$

$$=\Delta q_1^2(l_1^2+l_2^2)+2l_1l_2\Delta q_1^2\cos(q_2+\delta q_2)+2l_1l_2\Delta q_1\Delta q_2\cos(q_2+\delta q_2)+$$

$$2l_1l_2\Delta q_1\delta\Delta q_2^{r}\cos(q_2+\delta q_2)+l_2^2\Delta q_2^2+2l_2^2\Delta q_2\delta\Delta q_2^{r}+l_2^2\delta\Delta q_2^{r2}+$$

$$2l_2^2\Delta q_1\Delta q_2+2l_2^2\Delta q_1\delta\Delta q_2^{r}. \qquad (2.21)$$

We continue with the right side of $\Delta L_{initial}^2$ - ΔL_{final}^2 using (2.4) and taking into account that

$$\Delta L_{initial}^2 = [\Delta q_1^2(l_1^2+l_2^2)+ l_2^2\,\Delta q_2^2 + 2\Delta q_1\,\Delta q_2\,l_2^2\,] + 2\,l_1l_2(\Delta q_1^2+\Delta q_1\Delta q_2)\cos q_2. \qquad (2.22)$$

One can see that component $[\Delta q_1^2(l_1^2 + l_2^2)+ l_2^2\,\Delta q_2^2 + 2\Delta q_1\,\Delta q_2\,l_2^2\,]$ from $\Delta L_{initial}^2$ will disappear in $\Delta L_{initial}^2 - \Delta L_{final}^2$ because both $\Delta L_{final}^2(\delta\Delta q_2^{l})$ and $\Delta L_{final}^2(\delta\Delta q_2^{r})$ include the same component with opposite signs (see (2.20) and (2.21)). Using (2.20), (2.21) and (2.22), we can find then that:

$$\Delta L_{initial}^2 - \Delta L_{final}^2(\delta\Delta q_2^{l}) = 2\,l_1l_2(\Delta q_1^2 + \Delta q_1\Delta q_2)\cos q_2 - 2l_1l_2\Delta q_1^2\cos(q_2+\delta q_2)-$$

$$2l_1l_2\Delta q_1\Delta q_2\cos(q_2+\delta q_2)+\ 2l_1l_2\Delta q_1\delta\Delta q_2^{l}\cos(q_2+\delta q_2)+$$

$$2l_2^2\Delta q_2\delta\Delta q_2^{l}-l_2^2\delta\Delta q_2^{l2}+2l_2^2\Delta q_1\delta\Delta q_2^{l}$$

$$= -l_2^2\delta\Delta q_2^{l2}+\delta\Delta q_2^{l}\,2l_2(l_2(\Delta q_1+\Delta q_2)+l_1\Delta q_1\cos(q_2+\delta q_2))+$$

$$[2\,l_1l_2\Delta q_1\,((\Delta q_1 + \Delta q_2)\,(\cos q_2 - \cos(q_2+\delta q_2))]. \qquad (2.23)$$

$$\Delta L_{initial}^2 - \Delta L_{final}^2(\delta\Delta q_2^{r}) = 2\,l_1l_2(\Delta q_1^2 + \Delta q_1\Delta q_2)\cos q_2 - 2l_1l_2\Delta q_1^2\cos(q_2+\delta q_2)-$$

$$2l_1l_2\Delta q_1\Delta q_2\cos(q_2+\delta q_2)-\ 2l_1l_2\Delta q_1\delta\Delta q_2^{r}\cos(q_2+\delta q_2)-$$

$$2l_2^2\Delta q_2\delta\Delta q_2^{r}-l_2^2\delta\Delta q_2^{r2}-2l_2^2\Delta q_1\delta\Delta q_2^{r}$$

$$= -l_2^2 \delta\Delta q_2^{l^2} - \delta\Delta q_2^{l} 2 l_2 (l_2(\Delta q_1 + \Delta q_2) + l_1 \Delta q_1 \cos(q_2 + \delta q_2)) +$$

$$[2 l_1 l_2 \Delta q_1 ((\Delta q_1 + \Delta q_2) (\cos q_2 - \cos(q_2 + \delta q_2))]. \quad (2.24)$$

Using (2.19), we can present (2.23) and (2.24) as quadratic equations in the form:

$$A\, \delta\Delta q_2^{l^2} + B_1\, \delta\Delta q_2^{l} + C = 0$$

$$A\, \delta\Delta q_2^{r^2} + B_2\, \delta\Delta q_2^{r} + C = 0 \quad (2.25)$$

where

$A = -l_2^2$,

$B_1 = 2 l_2 (l_2(\Delta q_1 + \Delta q_2) + l_1 \Delta q_1 \cos(q_2 + \delta q_2))$,

$B_2 = -2 l_2 (l_2(\Delta q_1 + \Delta q_2) + l_1 \Delta q_1 \cos(q_2 + \delta q_2))$,

$C = [2 l_1 l_2 \Delta q_1 ((\Delta q_1 + \Delta q_2) (\cos q_2 - \cos(q_2 + \delta q_2))] - 2\Delta L_{initial}\, \delta L + \delta L^2$

Equations (2.25) represent the needed dependences $\delta\Delta q_2^{l}(\delta L, \delta q_2)$ and $\delta\Delta q_2^{r}(\delta L, \delta q_2)$. If solutions of quadratic equations (2.25) exist then they can be found as (Gmurman, 2000):

$$\delta\Delta q_2^{l} = \frac{-B_1 \pm \sqrt{B_1^2 - 4AC}}{2A}$$

$$\delta\Delta q_2^{r} = \frac{-B_2 \pm \sqrt{B_2^2 - 4AC}}{2A} . \quad (2.26)$$

The dependences $\delta L(\delta q_2, \delta\Delta q_2^{l})$ and $\delta L(\delta q_2, \delta\Delta q_2^{r})$ (see (2.17) and (2.18)) for 2-R robotic manipulator (*e.g.*, robot RM 10-01 with link lengths $l_1 = 0.4$ m and $l_2 = 0.25$ m), with various values of parameters (Δq_1, Δq_2, initial joint coordinates q_1 and q_2, joint error limits Δq_2^{l} and Δq_2^{r}, *etc.*) are shown in graphs in Fig. 2.15 – 2.18. The exemplary data for graphs in Fig. 2.15 – 2.18 is presented in Tables 2.15 – 2.18.

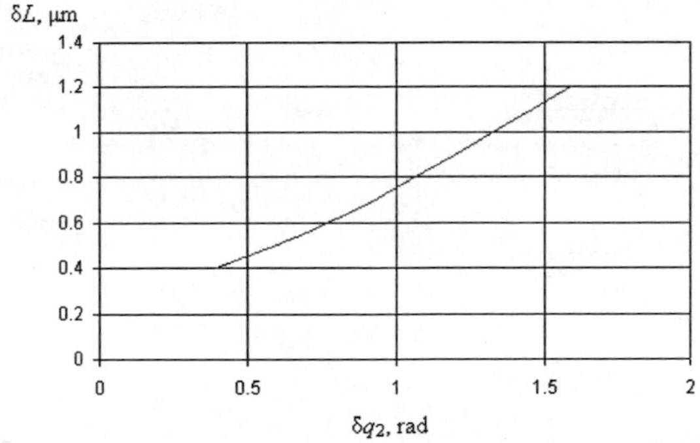

Fig. 2.15. Graph of $\delta L(\delta q_2, \delta\Delta q_2^{l})$ for industrial robot RM 10-01 ($\mathbf{P}_{initial}$ (q_1 = 1.77 rad; q_2 = 0.79 rad), l_1 = 0.4 m and l_2 = 0.25 m, $\Delta q_1 = 2.29 \times 10^{-6}$ rad, $\Delta q_2^{initial} = 2.21 \times 10^{-6}$ rad) with $\delta\Delta q_2^{l} = const = 1.00 \times 10^{-6}$ rad in the new point $\mathbf{P}_{final}$

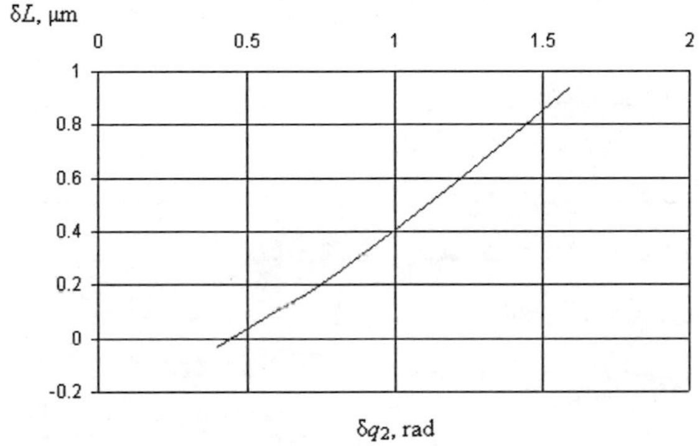

Fig. 2.16. Graph of $\delta L(\delta q_2, \delta\Delta q_2^{r})$ for industrial robot RM 10-01 ($\mathbf{P}_{initial}$ (q_1 = 1.77 rad; q_2 = 0.79 rad), l_1 = 0.4 m and l_2 = 0.25 m, $\Delta q_1 = 2.29 \times 10^{-6}$ rad, $\Delta q_2^{initial} = 2.21 \times 10^{-6}$ rad) with $\delta\Delta q_2^{r} = const = 1.00 \times 10^{-6}$ rad in the new point $\mathbf{P}_{final}$

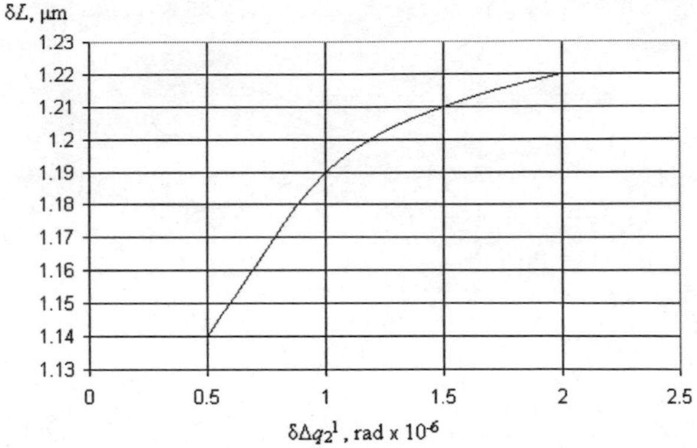

Fig. 2.17. Graph of $\delta L(\delta q_2, \delta \Delta q_2^{l})$ for industrial robot RM 10-01 ($\mathbf{P}_{initial}$ (q_1 = 1.77 rad; q_2 = 0.79 rad), l_1 = 0.4 m and l_2 = 0.25 m, $\Delta q_1 = 2.29 \times 10^{-6}$ rad, $\Delta q_2^{initial} = 2.21 \times 10^{-6}$ rad) with δq_2 = *const* = 1.57 rad in the new point $\mathbf{P}_{final}$

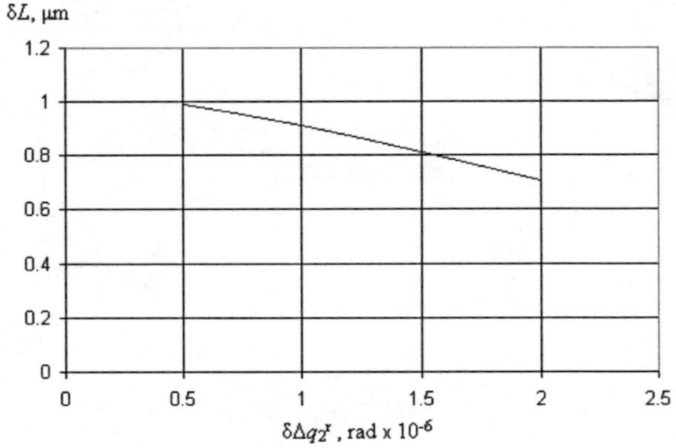

Fig. 2.18. Graph of $\delta L(\delta q_2, \delta \Delta q_2^{r})$ for industrial robot RM 10-01 ($\mathbf{P}_{initial}$ (q_1 = 1.77 rad; q_2 = 0.79 rad), l_1 = 0.4 m and l_2 = 0.25 m, $\Delta q_1 = 2.29 \times 10^{-6}$ rad, $\Delta q_2^{initial} = 2.21 \times 10^{-6}$ rad) with δq_2 = *const* = 1.57 rad in the new point $\mathbf{P}_{final}$

Table 2.15

Exemplary data for $\delta L(\delta q_2, \delta\Delta q_2^{l})$ of industrial robot RM 10-01 ($\mathbf{P}_{initial}$ (q_1 = 1.77 rad; q_2 = 0.79 rad), l_1 = 0.4 m and l_2 = 0.25 m, $\Delta q_1 = 2.29 \times 10^{-6}$ rad, $\Delta q_2^{initial} = 2.21 \times 10^{-6}$ rad) with $\delta\Delta q_2^{l} = const = 1.00 \times 10^{-6}$ rad in the new point $\mathbf{P}_{final}$

δq_2 [rad]	Final q_2 [rad]	$\Delta L_{initial}$ [μm]	ΔL_{final} [μm]	δL [μm]	Improvement / Deterioration [%]
0.4	1.19	1.86	1.46	0.40	21.4
0.8	1.59	1.86	1.24	0.62	34.5
1.2	1.99	1.86	0.96	0.9	48.2
1.6	2.39	1.86	0.65	1.21	65.0

Table 2.16

Exemplary data for $\delta L(\delta q_2, \delta\Delta q_2^{r})$ of industrial robot RM 10-01 ($\mathbf{P}_{initial}$ (q_1 = 1.77 rad; q_2 = 0.79 rad), l_1 = 0.4 m and l_2 = 0.25 m, $\Delta q_1 = 2.29 \times 10^{-6}$ rad, $\Delta q_2^{initial} = 2.21 \times 10^{-6}$ rad) with $\delta\Delta q_2^{r} = const = 1.00 \times 10^{-6}$ rad in the new point $\mathbf{P}_{final}$

δq_2 [rad]	Final q_2 [rad]	$\Delta L_{initial}$ [μm]	ΔL_{final} [μm]	δL [μm]	Improvement / Deterioration [%]
0.4	1.19	1.86	1.89	-0.03	-1.6
0.8	1.59	1.86	1.62	0.24	13.2
1.2	1.99	1.86	0.29	0.58	30.9
1.6	2.39	1.86	0.92	0.94	50.3

Table 2.17

Exemplary data for $\delta L(\delta q_2, \delta\Delta q_2^{l})$ of industrial robot RM 10-01 ($\mathbf{P}_{initial}$ (q_1 = 1.77 rad; q_2 = 0.79 rad), l_1 = 0.4 m and l_2 = 0.25 m, $\Delta q_1 = 2.29 \times 10^{-6}$ rad, $\Delta q_2^{initial} = 2.21 \times 10^{-6}$ rad) with δq_2 = *const* = 1.57 rad in the new point $\mathbf{P}_{final}$

$\delta\Delta q_2^{l}$ [rad] x 10^{-6}	Final q_2 [rad]	$\Delta L_{initial}$ [μm]	ΔL_{final} [μm]	δL [μm]	Improvement / Deterioration [%]
0.5	2.36	1.86	0.72	1.14	61.2
1	2.36	1.86	0.67	1.19	63.7
1.5	2.36	1.86	0.65	1.21	65.1
2.0	2.36	1.86	0.64	1.22	65.4

Table 2.18

Exemplary data for $\delta L(\delta q_2, \delta\Delta q_2^{r})$ of industrial robot RM 10-01 ($\mathbf{P}_{initial}$ ($q_1 = 1.77$ rad; $q_2 = 0.79$ rad), $l_1 = 0.4$ m and $l_2 = 0.25$ m, $\Delta q_1 = 2.29 \times 10^{-6}$ rad, $\Delta q_2^{initial} = 2.21 \times 10^{-6}$ rad) with $\delta q_2 = const = 1.57$ rad in the new point $\mathbf{P}_{final}$

$\delta\Delta q_2^{r}$ [rad] x 10^{-6}	Final q_2 [rad]	$\Delta L_{initial}$ [μm]	ΔL_{final} [μm]	δL [μm]	Improvement / Deterioration [%]
0.5	2.36	1.86	0.86	0.99	53.6
1	2.36	1.86	0.95	0.91	48.9
1.5	2.36	1.86	1.05	0.81	43.7
2.0	2.36	1.86	1.15	0.71	38.3
5.0	2.36	1.86	1.82	0.04	2.36

Graphs in Fig. 2.15 – 2.18 allow predicting usability of joint error maximum mutual compensation with worst case error approach for any 2-R planar robotic manipulator. As one can see from results presented in Tables 2.15 – 2.18, joint error maximum mutual compensation with worst case error approach will provide significant improvement (20 – 50 %) of end-effector positioning accuracy if joint error Δq_2 does not change by more than 50 – 80 % in the new point $\mathbf{P}_{final}$ (the exact values depend on the initial conditions, *i.e.*, values of δq_2, initial q_2 and other parameters (see (2.17) and (2.18) for details)). The probabilistic analysis performed in Chapter 2.2 for joint error maximum mutual compensation is also valid for worst case error approach.

We presented the worst case error approach for joint error maximum mutual compensation as applied to 2-R planar robotic manipulator. One has to follow the guidelines presented in Chapter 2.4, if one plans to implement the worst case error approach for more complex robotic manipulators (*e.g.*, 3-DOF, 6-DOF, *etc.*). It is also important to mention that based on the graph in Fig. 2.14, the effectiveness (predictability) of the joint error maximum mutual compensation becomes worse with the increase of the number of 2-R chains.

3. END-EFFECTOR POSITIONING ACCURACY IMPROVEMENT USING JOINT ERROR MAXIMUM MUTUAL COMPENSATION IN REDUNDANT ROBOTIC MANIPULATORS AND PUMA-560 TYPE ROBOTS

3.1. JOINT ERROR MAXIMUM MUTUAL COMPENSATION IN REDUNDANT ROBOTIC MANIPULATORS

A robotic manipulator is said to be redundant when it has more degrees of freedom than it is necessary for the specified task (Baker and Wampler, 1988; Chevallereau, 1998). The scheme of the 3-R redundant robotic manipulator self-motion in the given point of robot workspace is presented in Fig. 3.1. The extra degrees of freedom in the joint space provide an infinite number of inverse kinematic solutions at most joint configurations, lending the manipulator a proper self-motion. This can be utilized to determine the best redundant robotic manipulator joint configuration with joint error maximum mutual compensation. In the given research, the robotic redundancy was used to demonstrate the benefits of application of joint error maximum mutual compensation for redundant robotic manipulators. For example, the best manipulator joint configuration, that could provide better joint error maximum mutual compensation for a given working point in the robot workspace, can be found for a redundant robotic manipulator without changing the initial end-effector working point.

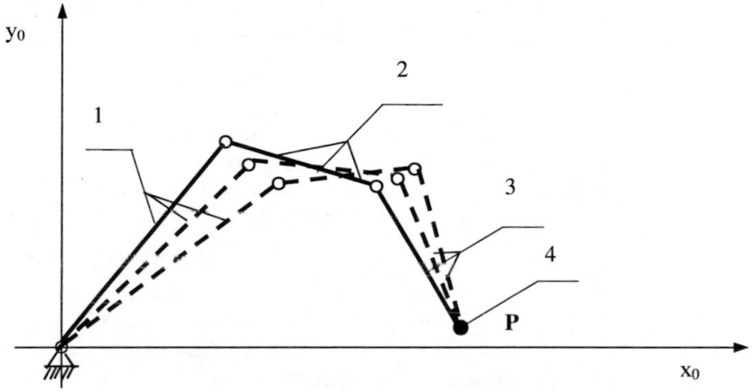

Fig. 3.1. Scheme of redundant robotic manipulator self-motion in the given point **P** of robot workspace (1 - first link, 2 - second link, 3 - third link, 4 - end-effector)

The application of joint error maximum mutual compensation for redundant robotic manipulators is similar to that of nonredundant robotic manipulators. This means that one can consider the redundant robotic manipulator as the chain of 2-R joints and perform joint error maximum mutual compensation for each of them as it was discussed in Chapter 2.4. Opposite to non-redundant robotic manipulators, one can do this without changing the initial working point of the robot.

For a redundant robotic manipulator, the Jacobian matrix $\mathbf{J}(\mathbf{q}) \in \mathbf{R}^{m \times n}$ ($m < n$ for redundant robotic manipulators; m is the number of Cartesian coordinates, n is the number of joint coordinates) defines the relation between robot end-effector errors $\Delta\mathbf{x} \in \mathbf{R}^m$ in base coordinate system and joint errors $\Delta\mathbf{q} \in \mathbf{R}^n$ as (Shahinpur, 1990):

$$\Delta\mathbf{x} = \mathbf{J}(\mathbf{q})\,\Delta\mathbf{q}.$$

In order to perform joint error maximum mutual compensation for a given end-effector location with coordinates $\mathbf{x} = const$ in the base coordinate system of the redundant robotic manipulator, one would look for a solution to the following problem:

$$\min(\Delta\mathbf{x})$$

subject to

$$\mathbf{x} = const$$

When applying joint error maximum mutual compensation to the redundant robotic manipulator, one has some benefits because of a possibility not to change the location of the end-effector working point of the robotic manipulator, but still obtain end-effector positioning accuracy improvement by changing robotic manipulator configuration in the given working point. To use this opportunity, the following method of end-effector positioning accuracy improvement using joint error maximum mutual compensation for redundant robotic manipulators has been proposed. Algorithm 3.1 of the proposed method for redundant robotic manipulators is presented below.

Algorithm 3.1

1. Determine a part of robot workspace in which the given technological operation is to be carried out.
2. Plan redundant robotic manipulator end-effector trajectory into the working point **P** using one of the trajectory planning methods, *e.g.*, (Bobrow, 1988; Bobrow, Dubowsky and Gibson, 1985; Dahl, 1994; Filonov, Kourtch and Veryha, 1999).
3. Determine redundant robotic manipulator configuration in the working point **P** with joint error maximum mutual compensation using modified method presented in Chapter 2.4. The modified method means that we also consider redundant robotic manipulator as composed of 2-R chains, however for the last joint of the redundant robotic manipulator we do not use joint error maximum mutual compensation. We simply calculate a joint coordinate that would position end-effector in initial point **P**. It is possible for redundant robotic manipulators due to their redundancy.
4. Plan new trajectory into the working point **P** with the new robot configuration found at step 3 and, thus, with maximum mutual compensation of joint errors.

The main difference and benefits of the presented approach for redundant robotic manipulators, comparing to nonredundant robotic manipulators, is that the position of the working point **P** (see Fig. 3.1) may not be changed to obtain end-effector positioning accuracy improvement. The manipulator should only change its configuration in the given point and plan a new trajectory into the new joint configuration (Filonov, Belajev and Veryha, 1999; Filonov and Veryha, 2000). Due to the fact that with the change of the robot joint configuration in the working point, the new trajectory will have to be planned, one should be sure that joint error values do not change significantly (it is empirically recommended to avoid joint error changes exceeding 1%) (Kieffer, Cahill and James, 1997). This condition should be taken into account for each particular case separately, ensuring that in the newly planned trajectory the new joint errors do not lead to an increase of the final end-effector positioning accuracy. Therefore, as soon as the new optimal robot joint configuration is identified in the given working point, one should check (if possible) end-effector accuracy characteristics in the given point while executing the planned trajectory into the new redundant robotic manipulator joint configuration.

3.2. KINEMATIC AND COMPUTER MODEL DETAILS OF 6-DOF PUMA - 560 TYPE ROBOTIC MANIPULATOR

PUMA-560 type robotic manipulator was used to perform computer simulation with the goal of searching for a theoretical maximum of end-effector positioning accuracy improvement using joint error maximum mutual compensation. The planned simulations should help us to make a conclusion about usability of joint error maximum mutual compensation for PUMA-560 type robotic manipulators. PUMA-560 is a 6-DOF robotic manipulator with revolute joints. The kinematics scheme of the PUMA - 560 type robotic manipulator is presented in Fig. 3.2. The joint coordinates have the following limits: q_1 (-160° ... 160°), q_2 (-225° ... 45°), q_3 (-225° ... 45°), q_4 (-110° ... 170°), q_5 (-100° ... 100°) and q_6 (-266° ... 266°) (Tchon, 1995). The geometry of PUMA-560 type robotic manipulator can be expressed using homogenous transformation matrix (Shahinpur, 1990; Tchon, 1995): $\begin{bmatrix} \mathbf{R}(\mathbf{q}) & \mathbf{T}(\mathbf{q}) \\ 0 & 1 \end{bmatrix}$, where $\mathbf{q}=(q_1, q_2, q_3, q_4, q_5, q_6)$ are joint coordinates (angles of displacement at each joint), $\mathbf{R}(\mathbf{q})$ is the matrix of rotations and $\mathbf{T}(\mathbf{q})$ is the matrix of translations.

The parameters of the PUMA - 560 type robotic manipulator according to the kinematic presentation model have the following values (Shahinpur, 1990) (see Fig. 3.2): $a_2 = 0.4138$ m, $a_3 = 0.02032$ m, $d_2 = 0.14909$ m, $d_4 = 0.43307$ m and $d_6 = 0.05625$ m. $\mathbf{R}(\mathbf{q})$ matrix of rotations for PUMA - 560 type robotic manipulator can be expressed as (Tchon, 1995):

$$\mathbf{R}(\mathbf{q}) = \begin{bmatrix} b_1c_1 - b_4s_1 & b_2c_1 - b_5s_1 & b_3c_1 - b_6s_1 \\ b_1s_1 - b_4c_1 & b_2s_1 - b_5c_1 & b_3s_1 + b_6c_1 \\ b_7 & b_8 & b_9 \end{bmatrix}, \tag{3.1}$$

where

$b_1 = c_{23}(c_4c_5c_6 - s_4s_6) - s_{23}s_5c_6,$

$b_2 = -c_{23}(c_4c_5s_6 + s_4c_6) + s_{23}s_5s_6,$

$b_3 = c_{23}c_4s_5 + s_{23}c_5,$

$b_4 = c_4s_6 + s_4c_5c_6,$

$b_5 = -c_4c_6 + s_4\, c_5\, s_6,$

$b_6 = s_4\, s_5,$

$b_7 = -\, s_{23}(c_4c_5c_6 - s_4s_6) - c_{23}s_5c_6,$

$b_8 = s_{23}(c_4c_5s_6 + s_4c_6) + c_{23}s_5s_6,$

$b_9 = -s_{23}c_4s_5 + c_{23}c_5.$

and following standard conventions from (Tchon, 1995):

$s_i = \sin q_i$, $c_i = \cos q_i$, $s_{ij} = \sin(q_i + q_j)$ and $c_{ij} = \cos(q_i + q_j)$.

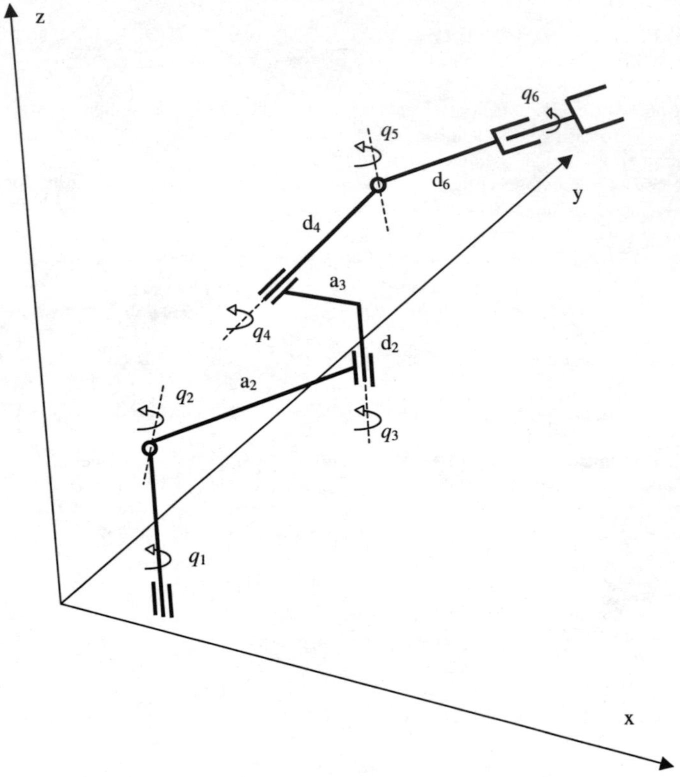

Fig. 3.2. PUMA - 560 type robotic manipulator kinematics

T(q) matrix of translations can be expressed as (Tchon, 1995):

$$\mathbf{T}(\mathbf{q}) = \begin{bmatrix} (c_{23}b_{10} + s_{23}b_{11} + a_2c_2)c_1 - b_{12}s_1 \\ (c_{23}b_{10} + s_{23}b_{11} + a_2c_2)s_1 + b_{12}c_1 \\ (-s_{23}b_{10} + c_{23}b_{11} - a_2s_2) \end{bmatrix}, \tag{3.2}$$

where

$b_{10} = d_6c_4s_5 + a_3,$

$b_{11} = d_6c_5 + d_4,$

$b_{12} = d_6s_4s_5 + d_2.$

PUMA - 560 type robotic manipulator Jacobian can be defined as (Tchon, 1995):

$$\mathbf{J}(\mathbf{q}) = \begin{bmatrix} j_{11} & j_{12} & j_{13} & j_{14} & j_{15} & 0 \\ j_{21} & j_{22} & j_{23} & j_{24} & j_{25} & 0 \\ 0 & j_{32} & j_{33} & j_{34} & j_{35} & 0 \\ 0 & -s_1 & -s_1 & c_1s_{23} & j_{45} & j_{46} \\ 0 & c_1 & c_1 & s_1s_{23} & j_{55} & j_{56} \\ 1 & 0 & 0 & c_{23} & s_{23}s_4 & j_{66} \end{bmatrix}, \tag{3.3}$$

where

$j_{11} = -(c_{23}b_{10} + s_{23}b_{11})s_1 - b_{12}c_1,$

$j_{21} = (c_{23}b_{10} + s_{23}b_{11})c_1 - b_{12}s_1,$

$j_{12} = (-s_{23}b_{10} + c_{23}b_{11} - a_2s_2)c_1,$

$j_{22} = (-s_{23}b_{10} + c_{23}b_{11} - a_2s_2)s_1,$

$j_{32} = -c_{23}b_{10} - s_{23}b_{11} - a_2c_2,$

$j_{13} = (-s_{23}b_{10} + c_{23}b_{11})c_{11},$

$j_{23} = (-s_{23}b_{10} + c_{23}b_{11})\, s_1,$

$j_{33} = -c_{23}b_{10} - s_{23}b_{11},$

$j_{14} = -d_6s_5(c_{23}s_4c_1 + c_4s_1),$

$j_{24} = -d_6s_5(c_{23}s_4s_1 - c_4c_1),$

$j_{34} = d_6s_{23}s_4s_5,$

$j_{15} = d_6((c_{23}c_4c_5 - s_{23}s_5)c_1 - s_4c_5s_1),$

$j_{25} = d_6((c_{23}c_4c_5 - s_{23}s_5)s_1 - s_4c_5c_1),$

$j_{35} = -d_6\,(s_{23}c_4c_5 + c_{23}s_5),$

$j_{45} = -c_4c_1 - c_{23}s_4c_1,$

$j_{55} = c_4c_1 - c_{23}\, s_4\, s_1,$

$j_{46} = -b_6s_1 + b_3c_1,$

$j_{56} = b_6c_1 + b_3s_1,$

$j_{66} = -c_{23}c_5 - s_{23}c_4s_5,$

and $b_1 \ldots b_{12}$ defined as in (3.1) and (3.2). The Cartesian positions of the end-effector (x, y, z) and orientation angles (φ, θ, ψ) can be used to define end-effector position and orientation in the base coordinate system. Positioning errors (Δx, Δy, Δz) and orientation errors ($\Delta\varphi$, $\Delta\theta$, $\Delta\psi$) can be expressed as (Tchon, 1995):

$$\begin{bmatrix} \Delta x \\ \Delta y \\ \Delta z \\ \Delta\varphi \\ \Delta\theta \\ \Delta\psi \end{bmatrix} = \begin{bmatrix} j_{11} & j_{12} & j_{13} & j_{14} & j_{15} & 0 \\ j_{21} & j_{22} & j_{23} & j_{24} & j_{25} & 0 \\ 0 & j_{32} & j_{33} & j_{34} & j_{35} & 0 \\ 0 & -s_1 & -s_1 & c_1 s_{23} & j_{45} & j_{46} \\ 0 & c_1 & c_1 & s_1 s_{23} & j_{55} & j_{56} \\ 1 & 0 & 0 & c_{23} & s_{23} s_4 & j_{66} \end{bmatrix} \begin{bmatrix} \Delta q_1 \\ \Delta q_2 \\ \Delta q_3 \\ \Delta q_4 \\ \Delta q_5 \\ \Delta q_6 \end{bmatrix}, \tag{3.4}$$

End-effector positioning accuracy $\Delta L = \sqrt{\Delta x^2 + \Delta y^2 + \Delta z^2}$ and end-effector orientation accuracy $\Delta L_a = \sqrt{\Delta\varphi^2 + \Delta\theta^2 + \Delta\psi^2}$ can be found for the PUMA - 560 type robotic manipulator based on (3.4). The presented kinematics of PUMA - 560 type robotic manipulator will be used in the next chapters to simulate end-effector positioning accuracy improvement using joint error maximum mutual compensation for 6-DOF robotic manipulators.

To perform computer simulation of end-effector positioning accuracy improvement of PUMA - 560 type robotic manipulator using joint error maximum mutual compensation, the following software tools were used and additionally developed:

- "ArmSol" (PUMA - 560 type robotic manipulator visualization tool developed at the Department of Robotic Systems of the Belarusian National Technical University for robotic manipulator motion visualization);
- "Accuracy Optimizer" (VBA (Visual Basic for Applications) tool for optimizing end-effector positioning accuracy that was developed as the part of this research for PUMA - 560 type robotic manipulator);
- MS Excel (standard Microsoft Office tool for generating reports).

The scheme of the integrated simulation framework is shown in Fig. 3.2. The simulation framework was specially developed to simplify the implementation of joint error maximum mutual compensation for 6-DOF PUMA - 560 type robotic manipulator. The alternative could be to analyze joints of 6-DOF PUMA - 560 type robotic manipulator separately or develop optimization approach using for example MatLab software, as it was done for hexapod robot in Chapter 4.

The main simulation tool in the given integrated framework is the "Accuracy Optimizer". It implements the global search algorithm (Algorithm 3.2, see below) using VBA program code and provides a flexible integration with the MS Excel for generating reports. The screen shots of the GUI (Graphical User Interface) provided by the "Accuracy Optimizer" are shown in Fig. 3.4 - 3.5. On the first stage of working with the "Accuracy Optimizer", the initial position of the working point should be defined using Cartesian or joint coordinates; additionally, joint errors should be defined. At the second stage, one should set one of the following optimization schemes: positioning optimization, positioning optimization only in the XY plane, positioning optimization only in the YZ plane, *etc.*, and pre-defined maximum allowed distance from the initial robot working point in the robot workspace at which it is supposed that joint errors do not change significantly (for example, empirically found, by not more than 1%) in practice in order not to deteriorate the simulation results. The results of the optimization are imported into the Excel sheet to form the optimization report and build graphs.

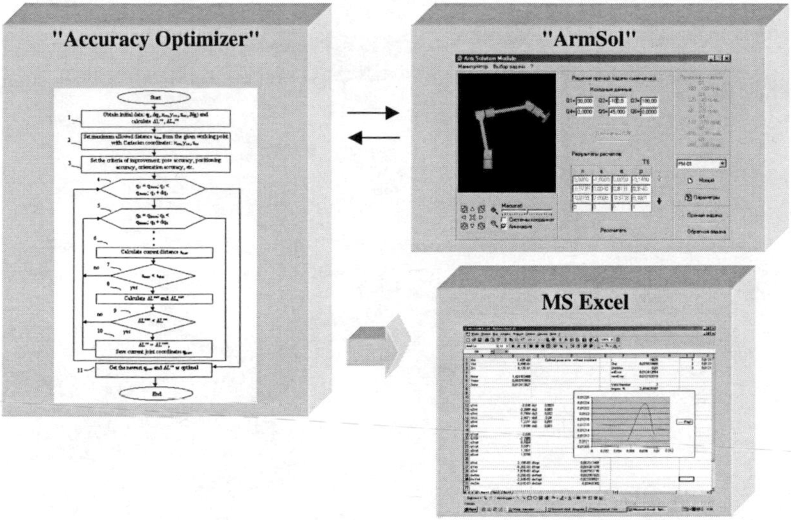

Fig. 3.3. The scheme of the integrated simulation framework for PUMA - 560 type robotic manipulator end-effector positioning accuracy improvement

The "ArmSol" tool is used to visualize PUMA - 560 type robotic manipulator configuration in the given working point. One can see the real position and configuration of the PUMA - 560 type robotic manipulator in the given working point, as well as easily transform Cartesian coordinates into the joint coordinates and backwards (solving either direct or inverse kinematic problem for robotic manipulator). The simulation framework with various known values of joint errors allowed performing the assessment of end-effector positioning accuracy improvement using joint error maximum mutual compensation based on the global search algorithm (Algorithm 3.2) presented below.

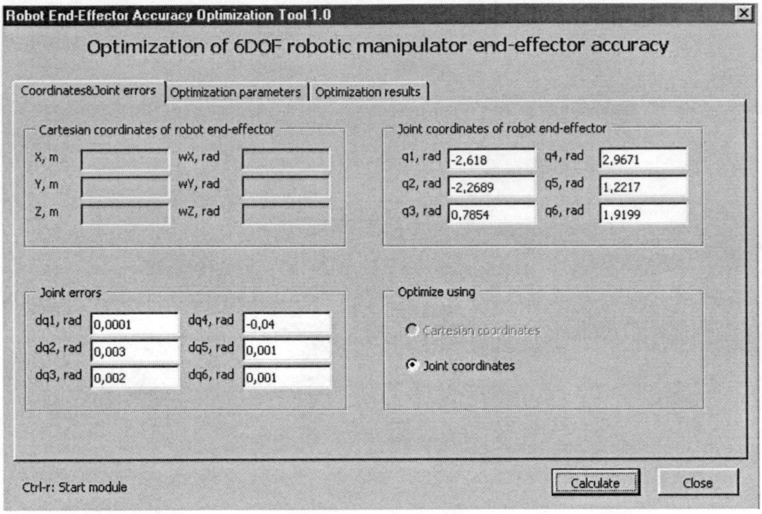

Fig. 3.4. The screen shot of "Accuracy Optimizer": Input of initial coordinates Tab

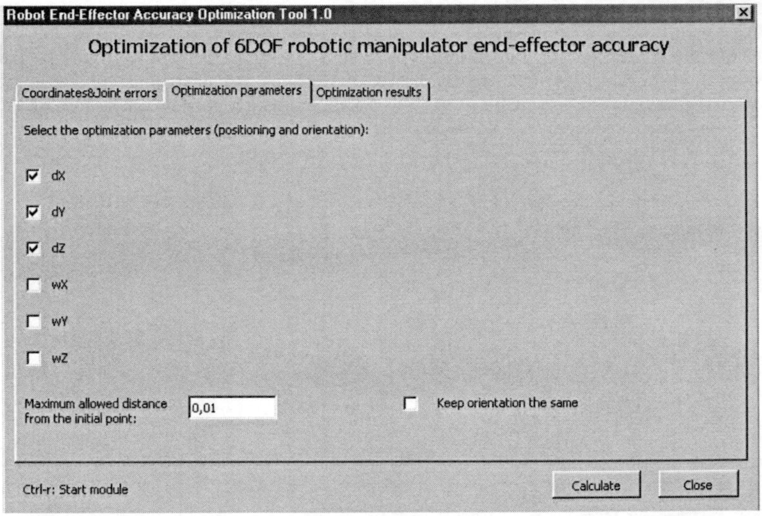

Fig. 3.5. The screen shot of "Accuracy Optimizer": Optimization parameters Tab

Algorithm 3.2

1. Obtain initial data for robot kinematic scheme: **q** (vector of joint coordinates), **Δq** (vector of joint errors), x_{ini}, y_{ini}, z_{ini} (Cartesian coordinates of initial point), **J**(**q**) (Jacobian matrix) and calculate ΔL^{ini} (initial end-effector positioning accuracy).
2. Set maximum allowed distance a_{dist} from the given working point with Cartesian coordinates: x_{ini}, y_{ini}, z_{ini}.
3. Set the criterion of improvement (*e.g.*, positioning accuracy, orientation accuracy, *etc.*).
4. Calculate end-effector positioning accuracy ΔL using in all possible end-effector positions within a_{dist} from initial point by changing joint coordinates with the given iteration step (the iteration continues by searching all possible coordinates with the given iteration steps $\delta q_1, \ldots, \delta q_n$ for n degree of freedom robotic manipulator).
5. Find minimum end-effector positioning accuracy **min(ΔL)** and joint coordinates in that position out of all obtained end-effector positioning accuracy **ΔL** values by comparing them with ΔL^{ini}.

Algorithm 3.2 is based on the manipulator Jacobian matrix that expresses the dependence of the linear and angular end-effector errors in the Cartesian coordinate system on joint errors (Shahinpur, 1990; Tchon, 1995):

$$\begin{pmatrix} \mathbf{\Delta L} \\ \mathbf{\Delta L}_{\mathrm{a}} \end{pmatrix} = \mathbf{J}(\mathbf{q})\mathbf{\Delta q}, \qquad (3.5)$$

where **J**(**q**) is the manipulator Jacobian matrix, **ΔL** is the vector of end-effector positioning accuracy, $\mathbf{\Delta L_a}$ is the vector of end-effector orientation accuracy and **Δq** is the vector of joint errors. To evaluate end-effector positioning accuracy for different manipulator configurations in the given robot workspace when joint error values are known, one may use (3.5), change joint coordinates in the manipulator Jacobian iteratively and calculate new values of end-effector positioning accuracy. As a result, one can easily simulate joint error maximum mutual compensation for robotic manipulators (see Chapter 2.4 for details) using presented approach to find maximum possible end-effector positioning accuracy improvement that can be obtained if joint error mutual compensation is applied in practice for PUMA - 560 type robotic manipulator.

3.3. COMPUTER SIMULATION OF PUMA - 560 ROBOT POSITIONING ACCURACY IMPROVEMENT USING JOINT ERROR MAXIMUM MUTUAL COMPENSATION

To investigate the effectiveness of joint error maximum mutual compensation for PUMA - 560 type robotic manipulator, the following computer simulations were performed in 8 points of the robot workspace. The working points $\mathbf{P}_1 \ldots \mathbf{P}_8$ were selected in 8 different parts of the base Cartesian coordinate system (based on (ISO 9283, 1998)), as it is shown in Fig. 3.6, to determine the influence of the location of the robot working point on the effectiveness of joint error maximum mutual compensation.

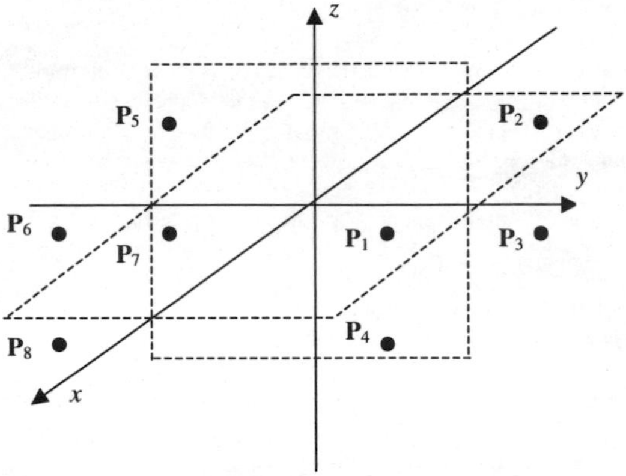

Fig. 3.6. Scheme of location of working points in base Cartesian coordinate system of PUMA - 560 type robotic manipulator

The joint coordinates of the robotic manipulator in selected points are presented in Table 3.1. The PUMA -560 type robotic manipulator positioning in the selected working points $\mathbf{P}_1 \ldots \mathbf{P}_8$ is shown in Fig. 3.7 and Fig. 3.8.

Table 3.1

Initial joint coordinates of working points

	$\mathbf{P}_1$	$\mathbf{P}_2$	$\mathbf{P}_3$	$\mathbf{P}_4$	$\mathbf{P}_5$	$\mathbf{P}_6$	$\mathbf{P}_7$	$\mathbf{P}_8$
q_1 [rad]	-2.618	-0.1745	-0.1745	-0.5236	0.5236	-2.0944	1.5708	2.0944
q_2 [rad]	-2.2689	-1.9199	-3.6652	0.6981	-1.9199	-0.8727	-3.6652	-3.6652
q_3 [rad]	0.7854	0.6981	0.6981	2.7925	0.6981	0.6981	0.6981	0.6981
q_4 [rad]	2.9671	0.5236	0.5236	0.5236	0.5236	0.5236	0.5236	0.5236
q_5 [rad]	1.2217	0.6981	0.6981	0.6981	0.6981	0.6981	0.6981	0.6981
q_6 [rad]	0.5236	0.5236	0.5236	0.5236	0.5236	0.5236	0.5236	0.5236

The improvement of end-effector positioning accuracy using joint error maximum mutual compensation was performed in the selected points $\mathbf{P}_1 \ldots \mathbf{P}_8$ according to the global search algorithm (Algorithm 3.2), using software tools "ArmSol" and "Accuracy Optimizer" on PC with 900 MHz and ~400 Mb RAM and assuming that joint errors had the following values in the initial and final points:

$$\Delta q_1 = 0.0001 \text{ rad}, \Delta q_2 = 0.003 \text{ rad}, \Delta q_3 = 0.002 \text{ rad},$$

$$\Delta q_4 = -0.04 \text{ rad}, \Delta q_5 = 0.001 \text{ rad and } \Delta q_6 = 0.001 \text{ rad}$$

and maximum allowed distance (for robot end-effector) a_{dist} from initial point was 0.01 m. Based on the global search algorithm (Algorithm 3.2) with the iteration of 0.001 rad for joint coordinates and an assumption that joint error values did not change within the end-effector location of 0.01 m from the initial point, the new PUMA - 560 robot working points were found with the improved end-effector positioning accuracy. The optimization results are presented in Tables 3.2 and 3.3.

Table 3.2

New joint coordinates of working points after optimization in points $\mathbf{P}_1 \ldots \mathbf{P}_8$

	$\mathbf{P}_1$	$\mathbf{P}_2$	$\mathbf{P}_3$	$\mathbf{P}_4$	$\mathbf{P}_5$	$\mathbf{P}_6$	$\mathbf{P}_7$	$\mathbf{P}_8$
q_1 [rad]	-2.628	-0.1845	-0.1845	-0.5136	0.5136	-2.0844	1.5708	2.0844
q_2 [rad]	-2.2589	-1.9099	-3.6552	0.6881	-1.9099	-0.8627	-3.6552	-3.6552
q_3 [rad]	0.7654	0.6781	0.6881	2.8025	0.6781	0.6781	0.6881	0.6881
q_4 [rad]	3.0071	0.4836	0.4836	0.4836	0.4836	0.4836	0.4836	0.4836
q_5 [rad]	1.1817	0.6581	0.6581	0.6581	0.6581	0.6581	0.6581	0.6581
q_6 [rad]	0.5236	0.5236	0.5236	0.5236	0.5236	0.5236	0.5236	0.5236

One should note that we did not set any limitations on the orientation of the robotic manipulator end-effector in the selected points, thus, PUMA – 560 type robot redundancy was available and would be automatically used by the Algorithm 3.2 if the use of redundancy could improve end-effector positioning accuracy during simulation. Iteration steps of 0.0001 rad and smaller would require much higher computer hardware performance to obtain the simulation results. High-performance servers were not available during the simulation tests.

Table 3.3

Simulation results of end-effector positioning accuracy improvement in points $\mathbf{P}_1 \ldots \mathbf{P}_8$

Point	Initial positioning accuracy [m]	New positioning accuracy [m]	Improvement, %	Distance from initial point [m]
$\mathbf{P}_1$	0.012412594	0.012103319	2.49	0.005530888
$\mathbf{P}_2$	0.012478478	0.012103713	3.01	0.007034213
$\mathbf{P}_3$	0.012451173	0.012086379	2.92	0.00484993
$\mathbf{P}_4$	0.010944128	0.010567304	3.44	0.00985531
$\mathbf{P}_5$	0.01239059	0.012019553	2.99	0.007034213
$\mathbf{P}_6$	0.012273371	0.011904208	3.02	0.009963912
$\mathbf{P}_7$	0.012391988	0.012030759	2.91	0.009711984
$\mathbf{P}_8$	0.01235064	0.01199156	2.90	0.00484993

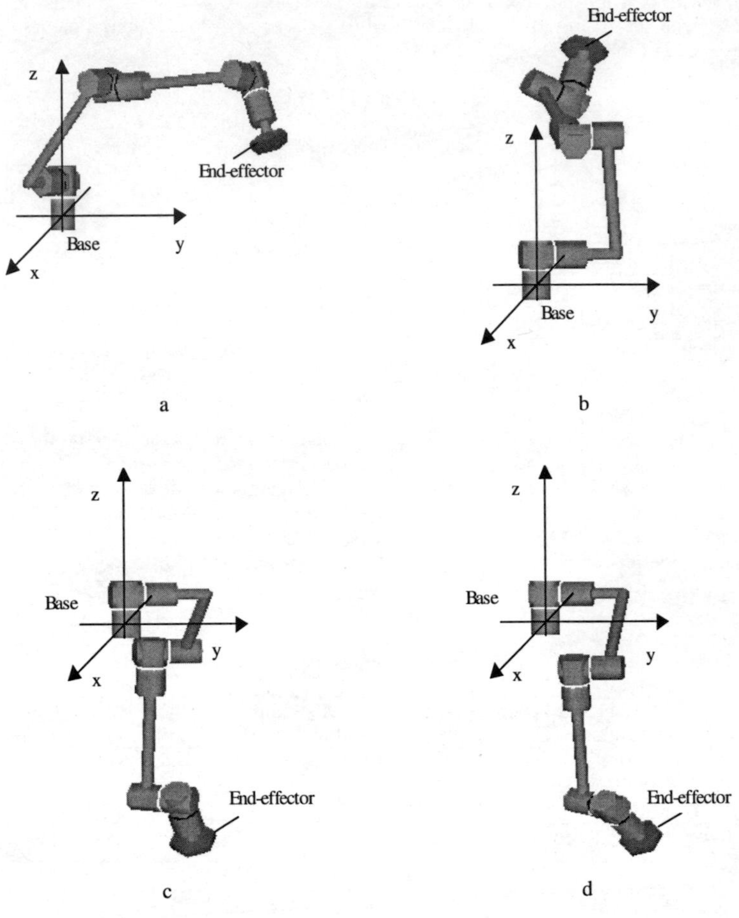

Fig. 3.7. Positioning of PUMA -560 type robotic manipulator in working points: a – in point $\mathbf{P}_1$; b – in point $\mathbf{P}_2$; c – in point $\mathbf{P}_3$; d - in point $\mathbf{P}_4$.

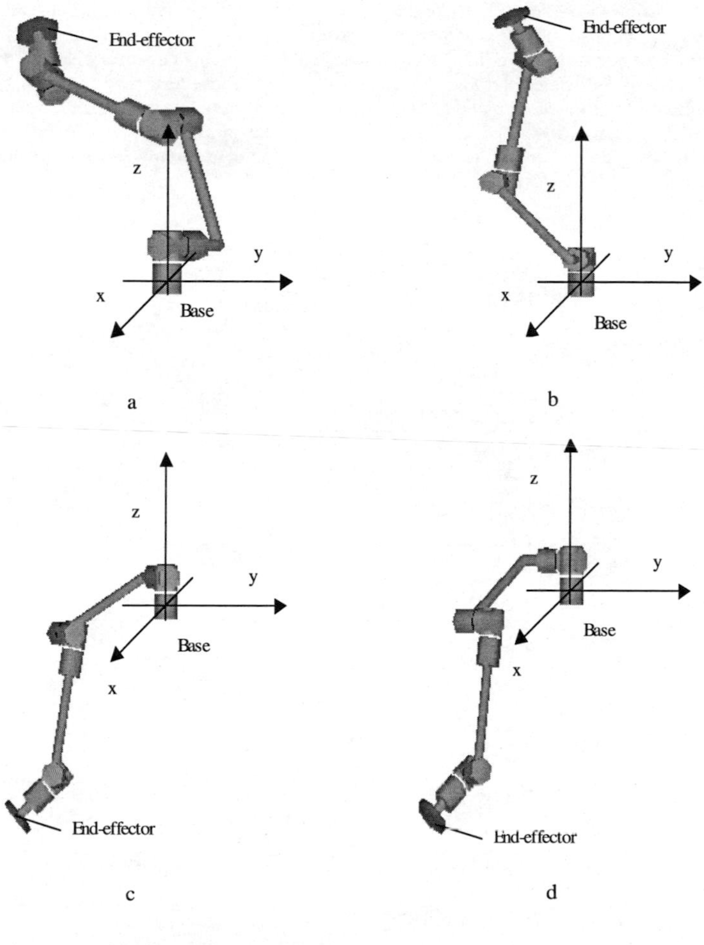

Fig. 3.8. Positioning of PUMA -560 type robotic manipulator in working points: a – in point $\mathbf{P}_5$; b – in point $\mathbf{P}_6$; c – in point $\mathbf{P}_7$; d – in point $\mathbf{P}_8$.

As it can be concluded based on simulation results in Table 3.3, in all examples, the robotic manipulator configuration changed only slightly and end-effector positioning accuracy was improved by average 3 % in the new robot working points because of the better joint error mutual compensation in these points. Hence, the performance of the method of joint error maximum mutual compensation depends only slightly on the location of the working point in the robot workspace.

To find out how each of joint coordinates (q_1, q_2, q_3, q_4, q_5 and q_6) separately contributed to the improved end-effector positioning accuracy during joint error mutual compensation (see Table 3.3), additional Tables 3.4 – 3.11 are presented. Tables 3.4 – 3.11 show the improvement of end-effector positioning accuracy if only one of the joint coordinates is changed at a time and all other joint coordinates remain unchanged. Graphs in Fig. 3.8 and 3.9 (based on data in Tables 3.4 – 3.11) show how many percent each joint coordinate contributed to the end-effector positioning accuracy improvement (see Table 3.3). Based on graphs in Fig. 3.8 and 3.9, one can see that in all working points $\mathbf{P}_1$ ... $\mathbf{P}_8$, the largest contribution to end-effector positioning accuracy improvement was from joint coordinates q_4 and q_5.

Table 3.4

End-effector positioning accuracy improvement in point $\mathbf{P}_1$ if only one joint coordinate is changed at a time

Joint coordinate	Joint coordinate change [rad]	End-effector positioning accuracy improvement, %
q_1	-0.01	~0
q_2	0.01	~0
q_3	-0.02	0.16
q_4	0.04	1.57
q_5	-0.04	0.76
q_6	0	0
		Σ = 2.49

Table 3.5

End-effector positioning accuracy improvement in point $\mathbf{P}_2$ if only one joint coordinate is changed at a time

Joint coordinate	Joint coordinate change [rad]	End-effector positioning accuracy improvement, %
q_1	-0.01	~0
q_2	0.01	~0
q_3	-0.02	0.2
q_4	-0.04	0.97
q_5	-0.04	1.84
q_6	0	0
		Σ = 3.01

Table 3.6

End-effector positioning accuracy improvement in point $\mathbf{P}_3$ if only one joint coordinate is changed at a time

Joint coordinate	Joint coordinate change [rad]	End-effector positioning accuracy improvement, %
q_1	-0.01	~0
q_2	0.01	~0
q_3	-0.01	0.1
q_4	-0.04	0.95
q_5	-0.04	1.87
q_6	0	0
		Σ = 2.92

Table 3.7

End-effector positioning accuracy improvement in point $\mathbf{P}_4$ if only one joint coordinate is changed at a time

Joint coordinate	Joint coordinate change [rad]	End-effector positioning accuracy improvement, %
q_1	0.01	~0
q_2	-0.01	~0
q_3	0.01	0.2
q_4	-0.04	1.07
q_5	-0.04	2.17
q_6	0	0
		Σ = 3.44

Table 3.8

End-effector positioning accuracy improvement in point $\mathbf{P}_5$ if only one joint coordinate is changed at a time

Joint coordinate	Joint coordinate change [rad]	End-effector positioning accuracy improvement, %
q_1	-0.01	~0
q_2	0.01	0.01
q_3	-0.02	0.17
q_4	-0.04	0.93
q_5	-0.04	1.88
q_6	0	0
		Σ = 2.99

Table 3.9

End-effector positioning accuracy improvement in point $\mathbf{P}_6$ if only one joint coordinate is changed at a time

Joint coordinate	Joint coordinate change [rad]	End-effector positioning accuracy improvement, %
q_1	0.01	~0
q_2	0.01	~0
q_3	-0.02	0.2
q_4	-0.04	0.92
q_5	-0.04	1.9
q_6	0	0
		Σ = 3.02

Table 3.10

End-effector positioning accuracy improvement in point $\mathbf{P}_7$ if only one joint coordinate is changed at a time

Joint coordinate	Joint coordinate change [rad]	End-effector positioning accuracy improvement, %
q_1	~0	~0
q_2	0.01	~0
q_3	-0.01	0.08
q_4	-0.04	0.95
q_5	-0.04	1.88
q_6	0	0
		Σ = 2.91

Table 3.11

End-effector positioning accuracy improvement in point $\mathbf{P}_8$ if only one joint coordinate is changed at a time

Joint coordinate	Joint coordinate change [rad]	End-effector positioning accuracy improvement, %
q_1	-0.01	~0
q_2	0.01	~0
q_3	-0.01	0.11
q_4	-0.04	0.92
q_5	-0.04	1.87
q_6	0	0
		Σ = 2.90

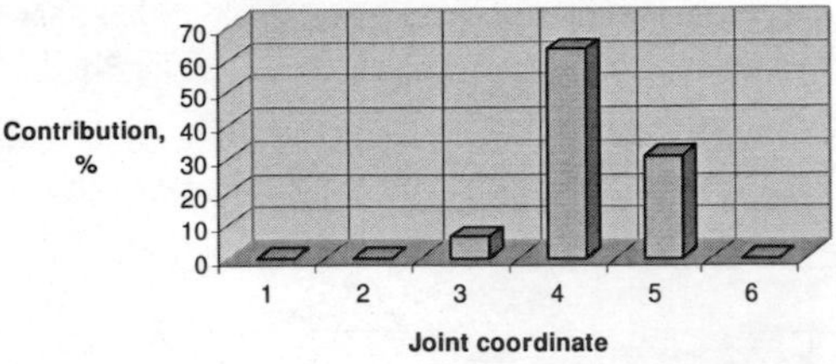

a

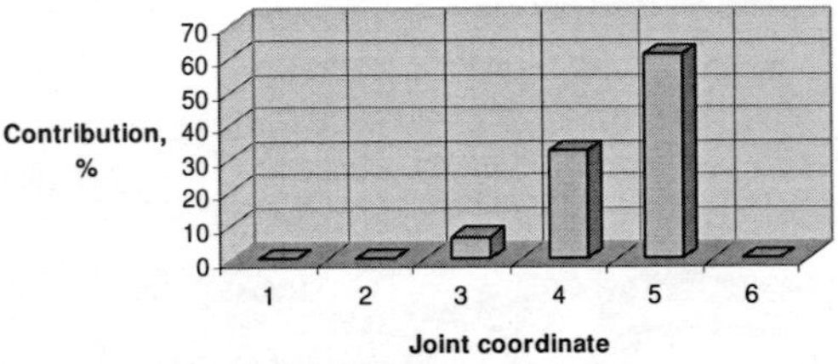

b

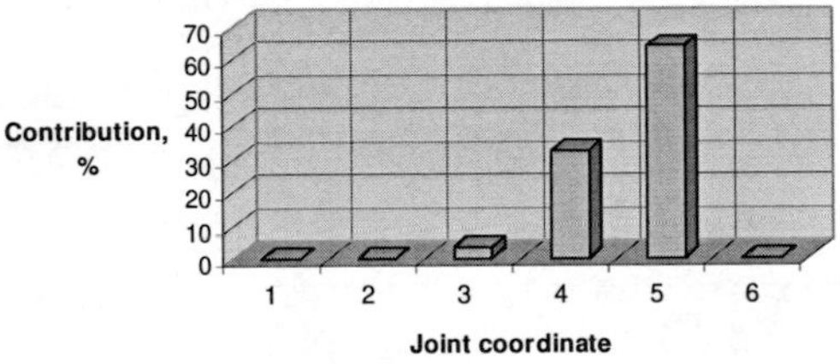

c

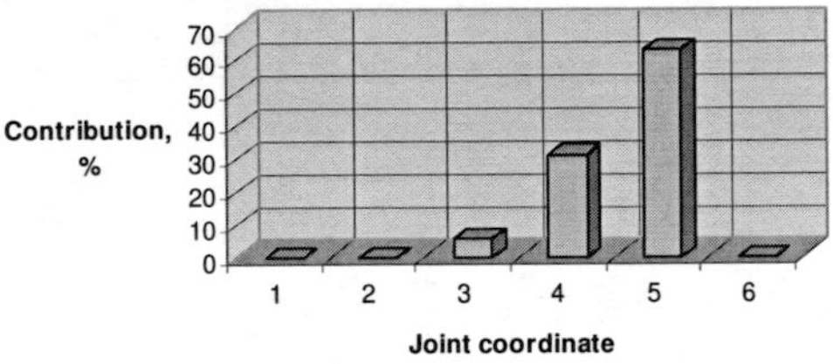

d

Fig. 3.8. Contribution of each joint coordinate to end-effector positioning accuracy improvement in percents: a – in point $\mathbf{P}_1$, b – in point $\mathbf{P}_2$, c – in point $\mathbf{P}_3$, d – in point $\mathbf{P}_4$

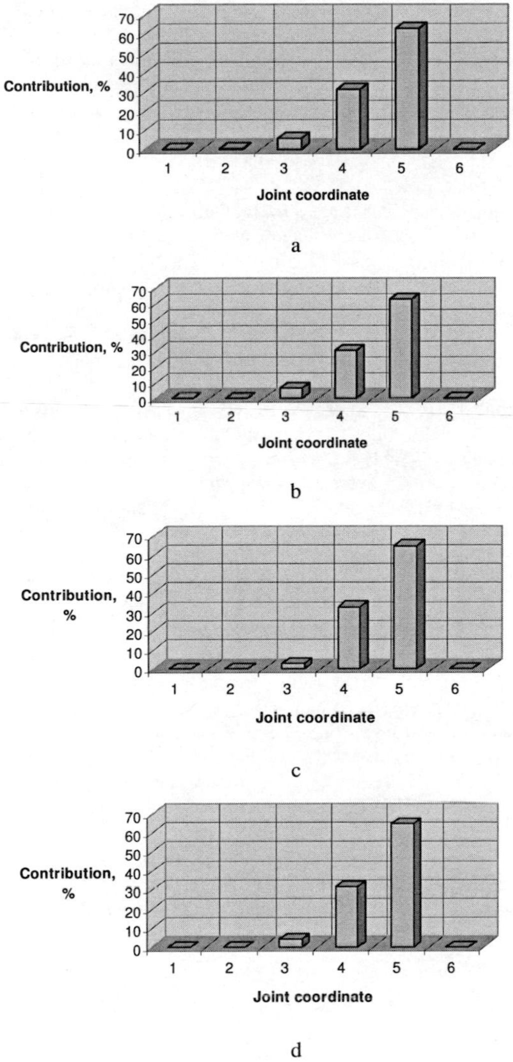

Fig. 3.9. Contribution of each joint coordinate to end-effector positioning accuracy improvement in percents: a – in point $\mathbf{P}_5$, b – in point $\mathbf{P}_6$, c – in point $\mathbf{P}_7$, d – in point $\mathbf{P}_8$

To investigate the influence of joint error signs and joint error absolute values on the effectiveness of joint error mutual compensation, additional simulations were performed by changing the values and signs of joint errors when manipulator positions itself in the given working point. The end-effector positioning accuracy improvement using joint error mutual compensation and simulation framework for the PUMA –560 type robotic manipulator was performed in the working point with joint coordinates q_1 = 0.5236 rad, q_2 = 0.7853 rad, q_3 = 2.7925 rad, q_4 = -0.1745 rad, q_5 = 0.8726 rad and q_6 = 0.7854 rad (Cartesian coordinates in the base system: x = -0.102485 m, y = 0.075643 m and z = -1.413604 m) and joint errors Δq_1 = 0.001 rad, Δq_2 = 0.001 rad, Δq_3 = 0.001 rad, Δq_4 = 0.001 rad, Δq_5 = 0.001 rad and Δq_6 = 0.001 rad. The maximum allowed distance a_{dist} for joint error maximum mutual compensation was a_{dist} = 0.01 m. The same simulation scheme was later used with different signs of joint errors. The results are presented in Table 3.12.

Based on the optimization results in Table 3.12 the graph showing the distribution of positioning accuracy improvement for different combinations of joint error signs of PUMA - 560 robot is presented in Fig. 3.10. As it is shown in graph in Fig. 3.10, the best combinations for joint error maximum mutual compensation of PUMA - 560 type robotic manipulator were number 5 and 28 (see Table 3.12). Under these combinations, the maximum value 1.25 % of the end-effector positioning accuracy improvement was obtained in the given example. The minimum value of 0.33 % of the end-effector positioning accuracy improvement was found for combinations 12 and 21 (see Table 3.12). Similar simulations can be performed for other working points and joint error values in the robot workspace.

The same simulations of the PUMA - 560 type robotic manipulator were performed to identify the influence of the change of joint error values on the improvement of PUMA - 560 robot positioning accuracy using joint error maximum mutual compensation. The end-effector positioning accuracy improvement using joint error maximum mutual compensation and simulation framework for the PUMA - 560 type robotic manipulator was performed in the working point with joint coordinates q_1 = 0.5236 rad, q_2 = 0.7853 rad, q_3 = 2.7925 rad, q_4 = -0.1745 rad, q_5 = 0.8726 rad and q_6 = 0.7854 rad (Cartesian coordinates in the base system: x = -0.102485 m, y = 0.075643 m and z = -1.413604 m) and joint errors Δq_1 = 0.001 rad, Δq_2 = 0.001 rad, Δq_3 = 0.001 rad, Δq_4 = 0.001 rad, Δq_5 = 0.001 rad and Δq_6 = 0.001 rad. The maximum allowed distance a_{dist} from the initial end-effector working point was 0.01 m. The simulation results are presented in Table 3.13. Based on the optimization results in Table 3.13, the graph presenting the distribution of positioning accuracy improvement for different combinations of joint errors of PUMA - 560 robot is shown in Fig. 3.11.

Table 3.12

Simulation results of PUMA -560 robot end-effector positioning accuracy improvement with different joint error signs (q_6 does not influence on simulation results (Tchon, 1995))

Test	Joint error [rad]					Positioning accuracy improvement, %
	Δq_1	Δq_2	Δq_3	Δq_4	Δq_5	
1	0.001	0.001	0.001	0.001	0.001	0.94
2	0.001	0.001	0.001	0.001	-0.001	0.55
3	0.001	0.001	0.001	-0.001	0.001	1.12
4	0.001	0.001	0.001	-0.001	-0.001	0.48
5	**0.001**	**0.001**	**-0.001**	**0.001**	**0.001**	**1.25**
6	0.001	0.001	-0.001	0.001	-0.001	0.41
7	0.001	0.001	-0.001	-0.001	0.001	0.78
8	0.001	0.001	-0.001	-0.001	-0.001	0.72
9	0.001	-0.001	0.001	0.001	0.001	0.45
10	0.001	-0.001	0.001	0.001	-0.001	1.11
11	0.001	-0.001	0.001	-0.001	0.001	0.88
12	0.001	-0.001	0.001	-0.001	-0.001	0.33
13	0.001	-0.001	-0.001	0.001	0.001	0.60
14	0.001	-0.001	-0.001	0.001	-0.001	1.04
15	0.001	-0.001	-0.001	-0.001	0.001	0.38
16	0.001	-0.001	-0.001	-0.001	-0.001	0.71
17	-0.001	0.001	0.001	0.001	0.001	0.71
18	-0.001	0.001	0.001	0.001	-0.001	0.38
19	-0.001	0.001	0.001	-0.001	0.001	1.04
20	-0.001	0.001	0.001	-0.001	-0.001	0.60
21	-0.001	0.001	-0.001	0.001	0.001	0.33
22	-0.001	0.001	-0.001	0.001	-0.001	0.88
23	-0.001	0.001	-0.001	-0.001	0.001	1.11
24	-0.001	0.001	-0.001	-0.001	-0.001	0.45
25	-0.001	-0.001	0.001	0.001	0.001	0.72
26	-0.001	-0.001	0.001	0.001	-0.001	0.78
27	-0.001	-0.001	0.001	-0.001	0.001	0.41
28	**-0.001**	**-0.001**	**0.001**	**-0.001**	**-0.001**	**1.25**
29	-0.001	-0.001	-0.001	0.001	0.001	0.48
30	-0.001	-0.001	-0.001	0.001	-0.001	1.12
31	-0.001	-0.001	-0.001	-0.001	0.001	0.55
32	-0.001	-0.001	-0.001	-0.001	-0.001	0.94

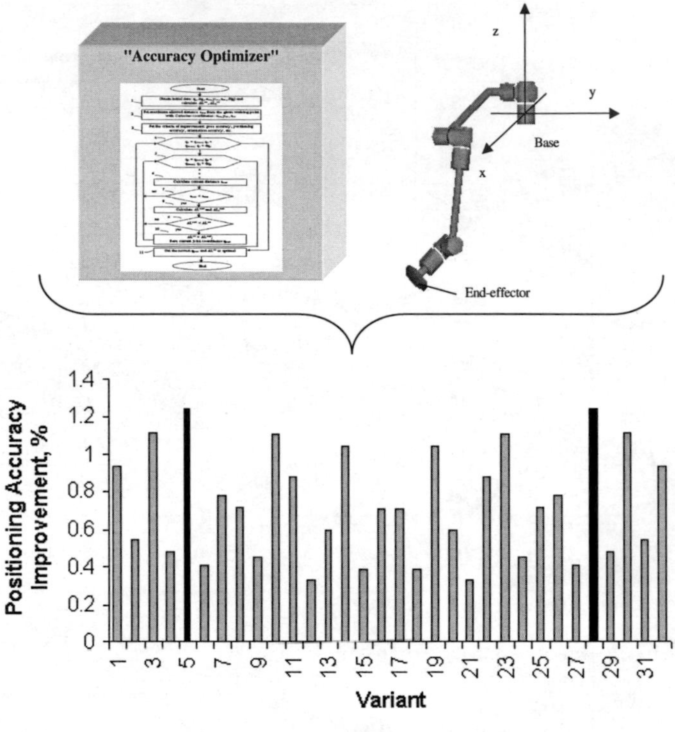

Fig. 3.10. Graph of distribution of positioning accuracy improvement for different combinations of joint error signs of PUMA - 560 robot

Table 3.13

Simulation results of PUMA - 560 robot end-effector positioning accuracy improvement with different joint error values

Test	Joint error [rad]					Positioning accuracy improvement, %
	Δq_1	Δq_2	Δq_3	Δq_4	Δq_5	
1	0.001	0.001	0.001	0.001	0.001	0.94
2	**0.001**	**0.001**	**0.001**	**0.001**	**0.002**	**1.16**
3	0.001	0.001	0.001	0.002	0.001	0.88
4	0.001	0.001	0.001	0.002	0.002	1.01
5	0.001	0.001	0.002	0.001	0.001	0.81
6	0.001	0.001	0.002	0.001	0.002	0.96
7	0.001	0.001	0.002	0.002	0.001	0.77
8	0.001	0.001	0.002	0.002	0.002	0.85
9	0.001	0.002	0.001	0.001	0.001	0.89
10	0.001	0.002	0.001	0.001	0.002	1.07
11	0.001	0.002	0.001	0.002	0.001	0.85
12	0.001	0.002	0.001	0.002	0.002	0.98
13	0.001	0.002	0.002	0.001	0.001	0.83
14	0.001	0.002	0.002	0.001	0.002	0.96
15	0.001	0.002	0.002	0.002	0.001	0.80
16	0.001	0.002	0.002	0.002	0.002	0.89
17	0.002	0.001	0.001	0.001	0.001	0.96
18	0.002	0.001	0.001	0.001	0.002	1.15
19	0.002	0.001	0.001	0.002	0.001	0.90
20	0.002	0.001	0.001	0.002	0.002	1.10
21	0.002	0.001	0.002	0.001	0.001	0.86
22	0.002	0.001	0.002	0.001	0.002	1.00
23	0.002	0.001	0.002	0.002	0.001	0.79
24	0.002	0.001	0.002	0.002	0.002	0.93
25	0.002	0.002	0.001	0.001	0.001	0.92
26	0.002	0.002	0.001	0.001	0.002	1.1
27	0.002	0.002	0.001	0.002	0.001	0.86
28	0.002	0.002	0.001	0.002	0.002	1.03
29	0.002	0.002	0.002	0.001	0.001	0.87
30	0.002	0.002	0.002	0.001	0.002	1.00
31	0.002	0.002	0.002	0.002	0.001	0.81
32	0.002	0.002	0.002	0.002	0.002	0.94

As it is shown in Fig. 3.11, the best combination for joint error maximum compensation of PUMA - 560 type robotic manipulator was number 2 (see Table 3.13). Under this combination, the maximum value 1.16 % of the end-effector positioning accuracy improvement was obtained in the given example. The minimum value of 0.77 % of the end-effector positioning accuracy improvement was found for combination 7 (see Table 3.13).

Similar simulations can be performed for other working points and joint error values in the robot workspace. As a result, special maps for end-effector positioning accuracy can be built for the given robotic manipulators to have an overview of the maximum of possible positive effect from joint error mutual compensation for different joint error values and signs.

These results can be later used in practice to make a decision if joint error maximum mutual compensation is useful to be implemented in the given case for a given robotic manipulator or not. The topic of building such special accuracy maps is outside the scope of this book.

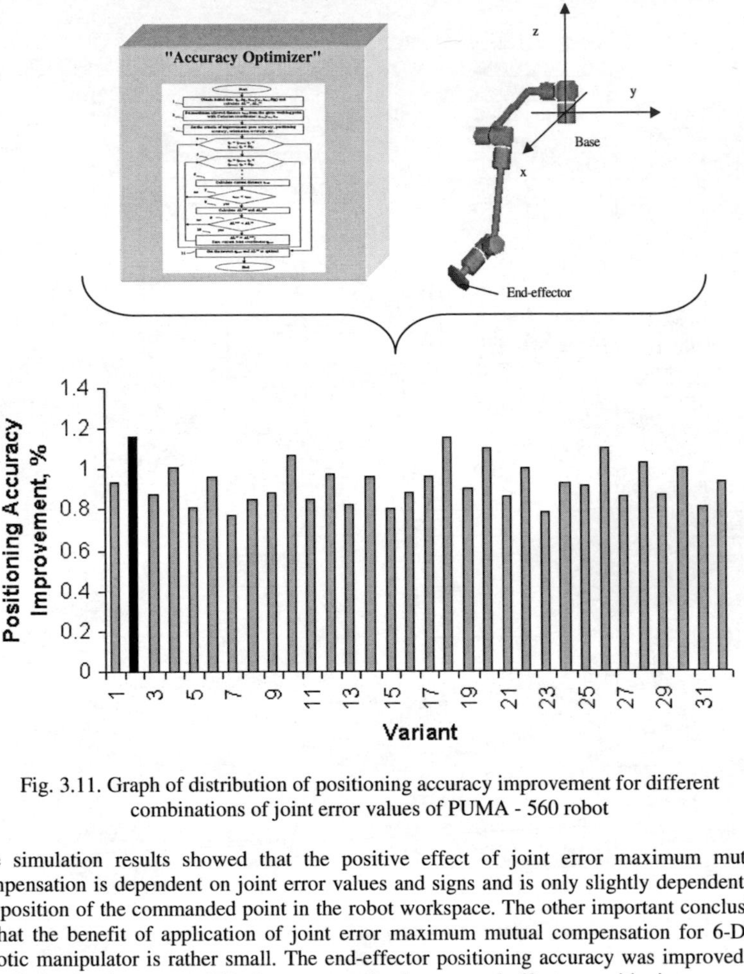

Fig. 3.11. Graph of distribution of positioning accuracy improvement for different combinations of joint error values of PUMA - 560 robot

The simulation results showed that the positive effect of joint error maximum mutual compensation is dependent on joint error values and signs and is only slightly dependent on the position of the commanded point in the robot workspace. The other important conclusion is that the benefit of application of joint error maximum mutual compensation for 6-DOF robotic manipulator is rather small. The end-effector positioning accuracy was improved by 3.5 % in the best case. The largest contribution to end-effector positioning accuracy improvement was observed from joint coordinates q_4 and q_5, when their values were decreased in PUMA-560 type robotic manipulator.

4. APPLICATION OF JOINT ERROR MAXIMUM MUTUAL COMPENSATION FOR HEXAPOD ROBOT

4.1. JOINT ERROR MAXIMUM MUTUAL COMPENSATION FOR HEXAPOD ROBOT

In this chapter, we present an approach for application of joint error maximum mutual compensation for positioning tasks in non-serial robotic manipulators. As an example, we selected a hexapod robot for our research (see (Doering, 2004) and Chapters 1.3, 5 for more details about Microbotic platform that uses hexapod robot for micro-handling). The hexapod robot belongs to the family of parallel robotic manipulators and has a relatively complex kinematics (see Fig. 4.1) (Nguyen, Zhou and Antrazi, 1991). Hence, we do not plan to try to find any analytical solution for it, as it was the case for robotic manipulators composed of 2-R joints (see Chapter 2.2). We will go for a numerical solution to find the optimal hexapod configurations with joint error maximum mutual compensation in which the positioning accuracy of hexapod end-effector (*i.e.*, triangle plate in Fig. 4.1) can be the best.

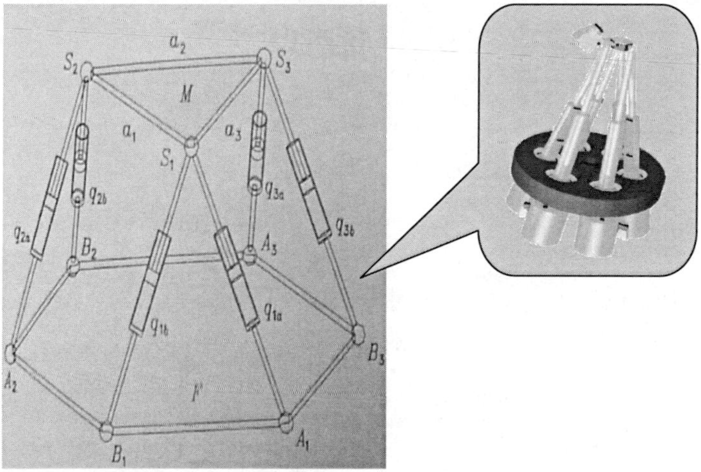

Fig. 4.1. Hexapod robotic manipulator and its 3-D model

We propose to find optimal hexapod configurations with joint error maximum mutual compensation for positioning of hexapod end-effector using the following approach. First, six joint coordinates **q** of hexapod actuator legs (Angeles, 2003) can be found as:

$$\mathbf{q} = f(\mathbf{x}), \tag{4.1}$$

where **x** is the vector containing translation and rotation parameters of hexapod end-effector. Second, joint errors $\Delta\mathbf{q}$ (Δq_1, Δq_2, Δq_3, Δq_4, Δq_5, Δq_6) in actuator legs of hexapod robot can be presented as (Angeles, 2003):

$$\Delta\mathbf{q} = \mathbf{J}\,\Delta\mathbf{x}, \tag{4.2}$$

where **J** is the Jacobian of the hexapod robotic manipulator and $\Delta\mathbf{x}$ (Δx, Δy, Δz, $\Delta\varphi$, $\Delta\theta$, $\Delta\psi$) are end-effector errors. As you can see, this expression of joint errors $\Delta\mathbf{q}$ using end-effector errors $\Delta\mathbf{x}$ looks different from that for serial robotic manipulators, where Jacobian is located on the left side of equation (4.2). Error vectors $\Delta\mathbf{x}$, $\Delta\mathbf{q}$ and Jacobian matrix **J** could be also normalized or scaled to appropriate units to enable easier comparisons between positioning errors (Δx, Δy, Δz) and orientation errors ($\Delta\varphi$, $\Delta\theta$, $\Delta\psi$) and to better estimate their effect on overall hexapod pose accuracy (Jelenkovic and Budin, 2002). Using (4.2), we can express $\Delta\mathbf{x}$ as:

$$\Delta\mathbf{x} = \mathbf{J}^{-1}\,\Delta\mathbf{q}\,. \tag{4.3}$$

Third, let us consider that unknown joint errors are in the interval (-ε, ε), which means $|\,\Delta\mathbf{q}\,| \leq \varepsilon$. Taking into account that infinity norm $\|\,\Delta\mathbf{q}\,\|_{\infty} = \max_{i=1..6} |\,\Delta\mathbf{q}_i\,|$ (i = 1 .. 6, because we have six actuators in hexapod), we can write that ΔL_{max} (worst-case positioning accuracy of hexapod end-effector under the given joint errors) can be found as:

$$\Delta L_{max} = \max\left(_{\|\Delta\mathbf{q}\|_{\infty} \leq \varepsilon} \{\|\,\Delta\mathbf{x}\,\|\}\right). \tag{4.4}$$

To find the optimal configuration of hexapod robot with joint error maximum mutual compensation and best positioning accuracy, our task is to find the minimum of ΔL_{max}. We continue with equation (4.3) and can write by definition (Chong and Zak, 2001) that:

$$\|\,\Delta\mathbf{x}\,\|_{\infty} \leq \|\,\mathbf{J}^{-1}\,\|_{\infty}\,\|\,\Delta\mathbf{q}\,\|_{\infty}\,, \tag{4.5}$$

where $\|\,\Delta\mathbf{x}\,\|_{\infty}$ and $\|\,\mathbf{J}^{-1}\,\|_{\infty}$ are, accordingly, the infinity norms of Cartesian end-effector errors and inverse Jacobian. Taking into account that $\|\,\Delta\mathbf{q}\,\|_{\infty} \leq \varepsilon$ and taking $\varepsilon = 1$ for a simplicity, based on (4.4) and (4.5), one can easily find that the optimal configuration for positioning accuracy of hexapod end-effector can be found as:

$$\Delta L_{max} \approx \|\,\mathbf{J}^{-1}\,\|_{\infty}\,. \tag{4.6}$$

Thus, to find the minimum of ΔL_{max} we have to find the minimum of infinity norm $\|\,\mathbf{J}^{-1}\,\|_{\infty}$ and our computational optimization task in the next sub-chapters is to find:

$$\min\left(\|\,\mathbf{J}^{-1}\,\|_{\infty}\right). \tag{4.7}$$

To solve (4.7), we will have to consider the details of hexapod robot kinematics, which will be discussed in Chapter 4.2. Algorithm 4.1 provides the details of the general approach to find configurations with joint error maximum mutual compensation for hexapod robots.

Algorithm 4.1

1. Determine a part of robot workspace in which the given technological operation is to be carried out.
2. Determine the working point $\mathbf{P}_{initial}$ for hexapod end-effector in the given part of the robot workspace (usually, the choice of this point is limited by the location of technological equipment).
3. Determine, using (4.7) and one of the computational algorithms (*e.g.*, one of Quasi-Newton methods (Chong and Zak, 2001)), new point $\mathbf{P}_{final}$ for hexapod end-effector

(triangle plate), which is found as the local minimum of $\| \mathbf{J}^{-1} \|_{\infty}$ in the robot workspace with joint error maximum mutual compensation.

4. Plan a trajectory into the point $\mathbf{P}_{final}$ of hexapod end-effector so that the new configuration of the robotic manipulator with the better mutual compensation of joint errors and, thus, improved positioning accuracy, is used for a given technological operation.

4.2. KINEMATIC MODEL DETAILS OF HEXAPOD ROBOT

One of the possible approaches to study the geometry and kinematics of hexapod robotic manipulator is to consider its legs separately, as depicted in Fig. 4.2. The Denavit-Hartenberg parameters of the leg shown in Fig. 4.2 are given in Table 4.1 One can see from Fig. 4.2 that the hexapod leg is a decoupled manipulator (Angeles, 2003).

Table 4.1

Denavit-Hartenberg parameters of the hexapod leg (see Fig. 4.2)

i	a_i, [rad]	b_i, [m] or θ_i, [rad]	α_i, [rad]
1	0	0	$\pi/2$
2	0	0	$\pi/2$
3	0	b_3	0
4	0	b_4 = const	$\pi/2$
5	0	0	$\pi/2$
6	0	b_6 = const	0

The only active joint of such manipulator is the third joint with joint coordinate b_3 (see the callout in Fig. 4.2). As an example (see the callout in Fig. 4.2), we show that the inverse kinematics for a positioning problem to find joint coordinates θ_1, θ_2, b_3 in one of hexapod legs based on Cartesian coordinates of point C (x_C, y_C, z_C) can be solved as (Angeles, 2003):

$$\theta_1 = \mathrm{arctg}\,(y_C / x_C)\,, \quad (4.8)$$

$$\theta_2 = \mathrm{arctg}\,((-x_C \cos(\theta_1) - y_C \sin(\theta_1)) / z_C)\,, \quad (4.9)$$

$$b_3 = \sqrt{x_C^2 + y_C^2 + z_C^2} - b_4\,, \quad (4.10)$$

The direct kinematics for a positioning problem to find Cartesian coordinates of point C (x_C, y_C, z_C) based on joint coordinates (see the callout in Fig. 4.2) can be solved as (Angeles, 2003):

$$x_C = (b_3 + b_4) \sin(\theta_2) \cos(\theta_1), \quad (4.11)$$

$$y_C = (b_3 + b_4) \sin(\theta_2) \sin(\theta_1)\,, \quad (4.12)$$

$$z_C = -(b_3 + b_4) \cos(\theta_2)\,, \quad (4.13)$$

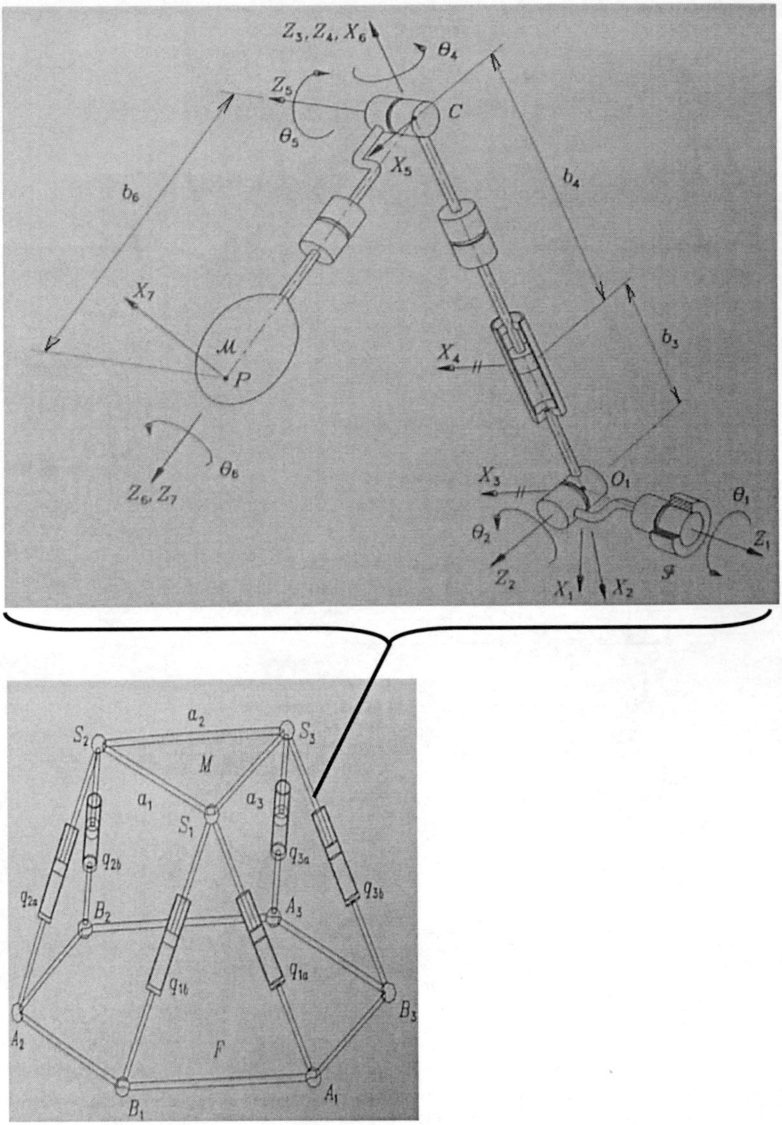

Fig. 4.2. Geometry of hexapod robot and one of its legs (Angeles, 2003)

To estimate the Jacobian matrix that relates the active motions of legs of the hexapod robot to the motion of the end-effector, one can consider the vectorial representation of the hexapod presented in Fig. 4.3 (Angeles, 2003). In Fig. 4.3, we used { B } for inertial reference frame of the base platform (assumed to be fixed) with center O, { P } for a reference frame at the center of mass C of the end-effector platform, $\bar{\mathbf{r}}_i$ is the vector that identifies the extremity of leg *i* in the base platform, $\bar{\mathbf{p}}_i$ is the vector that identifies the extremity of leg *i* in the end-effector platform, $\bar{\mathbf{x}}_o$ is the vector connecting the center O of the base platform with the center C of the end-effector platform and $\bar{\mathbf{l}}_i$ is the unit vector along leg *i*.

Let us consider that **R** is the rotation matrix, defined for example in terms of roll, pitch and yaw angles $\boldsymbol{\theta}$ (α, β, γ), relating { P } to { B }. The relationship between { P } and { B } is completely defined by $\mathbf{x}_0$ and $\boldsymbol{\theta}$. The Jacobian **J** relates the velocities of active joints $\dot{\mathbf{q}}$ to the velocity vector (twist of the end-effector platform at the operation point) $\dot{\mathbf{t}}$ ($\mathbf{v}^T$, $\boldsymbol{\omega}^T$), where $\mathbf{v} = \dot{\mathbf{x}}_0$ and $\boldsymbol{\omega} = \dot{\boldsymbol{\theta}}$, as:

$$\dot{\mathbf{q}} = \mathbf{J}\,\dot{\mathbf{t}}\,, \qquad (4.14)$$

The analytical expression can be obtained by expressing the absolute velocity $\mathbf{v}_i$ of the extremity $\mathbf{p}_i$ and projecting it along $\mathbf{l}_i$ to get:

$$\mathbf{v}_i = \mathbf{v} + \boldsymbol{\omega} \times \mathbf{p}_i\,,$$

where **v** is the absolute velocity of C and $\boldsymbol{\omega}$ is the angular velocity of the end-effector platform.

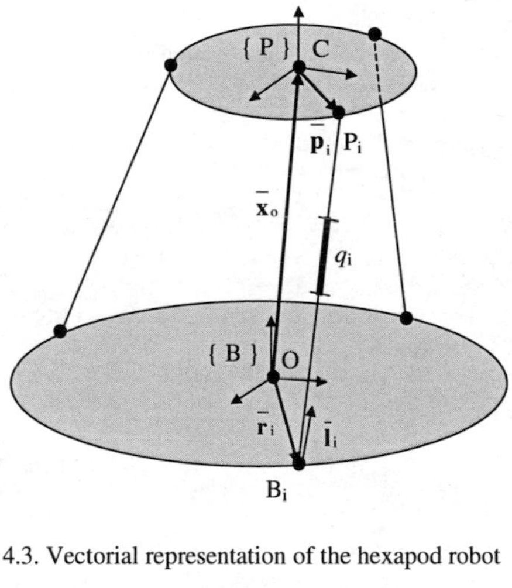

Fig. 4.3. Vectorial representation of the hexapod robot

After some transformations (see (Angeles, 2003) for details), one can obtain the final expression for Jacobian **J** as:

$$\mathbf{J} = \begin{pmatrix} \dots & \dots \\ \frac{1}{l_i}\left[(\mathbf{x}_0 - \mathbf{r}_i)^T \mathbf{R} + \mathbf{p}_i^T\right] & -\frac{1}{l_i}(\mathbf{x}_0 - \mathbf{r}_i)^T \mathbf{R}\tilde{\mathbf{p}}_i \\ \dots & \dots \end{pmatrix} \begin{pmatrix} \mathbf{v} \\ \boldsymbol{\omega} \end{pmatrix}, \tag{4.15}$$

where $\tilde{\mathbf{p}}_i$ is the antisymmetric matrix. It is important to mention that in our further discussions of joint error maximum mutual compensation for hexapod robot, we will consider only joint errors in actuators of hexapod legs (see (4.2)). Joint errors in passive joints of hexapod legs (see Fig. 4.2) can be readily found if joint errors in active joints are known (Angeles, 2003).

4.3. DESCRIPTION OF COMPUTER MODELS AND TOOLS FOR POSITIONING SIMULATION OF HEXAPOD ROBOT

To calculate minimum of infinity norm $\| \mathbf{J}^{-1} \|_\infty$ (see (4.7)), we will use MatLab software Release 14 with Service Pack 1 and Optimization Toolbox. MatLab is a high-performance language for technical computing. It integrates computation, visualization, and programming in an easy-to-use environment where problems and solutions are expressed in familiar mathematical notation. Typical use cases for MatLab include:

- Math and computation;
- Algorithm development;
- Data acquisition;
- Modeling, simulation and prototyping;
- Data analysis, exploration and visualization;
- Scientific and engineering graphics;
- Application development, including graphical user interface building.

To calculate minimum of the infinity norm $\| \mathbf{J}^{-1} \|_\infty$ and take into account limitations of hexapod robot workspace, the best choice would be to select one of the constrained optimization algorithms (Chong and Zak, 2001). However, the implementation of constrained optimization algorithms is in most cases very complex. Thus, we decided to limit our task and used unconstrained optimization algorithms. We have assessed available unconstrained optimization algorithms and their implementations (Chong and Zak, 2001). Based on our study, among quasi-Newton methods the BFGS (Broyden, Fletcher, Goldfarb and Shanno) algorithm is the most successful in published studies. It is one of the fastest and shows good convergence properties. Thus, we will use BFGS algorithm for solving our minimization task. Another positive aspect of using BFGS algorithm is that it has a stable implementation in MatLab Optimization Toolbox.

The visualization of hexapod robot in MatLab is shown in Fig. 4.4. The program listings in MatLab for calculating the minimum of infinity norm $\| \mathbf{J}^{-1} \|_\infty$ can be found in Appendix A.

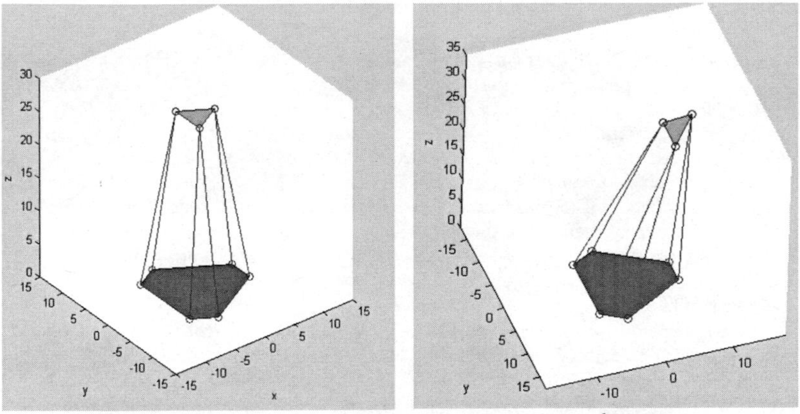

a b

Fig. 4.4. Visualization of hexapod robot in MatLab:
a – in neutral position and b – at the end of the trajectory

In practice, one can use the developed MatLab module as it is shown in the scheme in Fig. 4.5. The developed MatLab module for minimization of infinity norm $\| \mathbf{J}^{-1} \|_\infty$ can be accessed using the MatLab Application Program Interface (API) by the trajectory generator of robot control system. MatLab API is a library that allows writing C and Fortran programs that interact with MatLab.

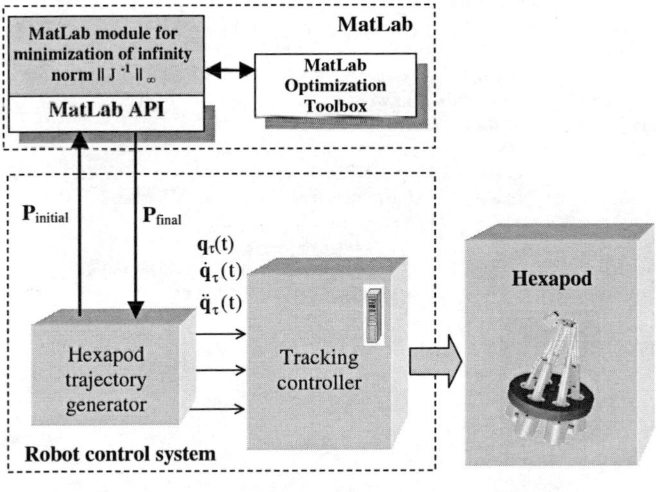

Fig. 4.5. Application of MatLab module for minimization
of infinity norm $\| \mathbf{J}^{-1} \|_\infty$ in test lab

MatLab API includes facilities for calling routines from MatLab (dynamic linking), calling MatLab as a computational engine, and for reading and writing MAT-files. As it was mentioned in Algorithm 4.1, hexapod trajectory generator will provide the coordinates (translation and rotation parameters) of working point $\mathbf{P}_{initial}$ of hexapod end-effector for a given technological operation to the developed MatLab module to find the improved position $\mathbf{P}_{final}$ based on minimization of infinity norm $\| \mathbf{J}^{-1} \|_{\infty}$. The MatLab module will minimize $\| \mathbf{J}^{-1} \|_{\infty}$ and return coordinates of the new end-effector point $\mathbf{P}_{final}$, in which better end-effector positioning accuracy can be expected because of the improved joint error mutual compensation. Hexapod trajectory generator can plan then a new trajectory into the newly found $\mathbf{P}_{final}$, if allowed under the given circumstances (*e.g.*, limitations from technological equipment, robot workspace, *etc.*).

Unfortunately, our solution in MatLab can not provide real-time calculations for hexapod trajectory corrections to improve end-effector positioning accuracy. One would require the development of the same optimization functionality in C or C++, which would be very time-consuming taking into account that one has to develop a very stable BFGS method implementation in C++ with:

- handling of very complex functions for which one cannot easily obtain gradients (*e.g.*, in MatLab one can use a built-in method of finite differences to approximate the function gradient);
- support for minimization of functions, which include such complex operations as inversion of matrices, infinity matrix norms, *etc.*

We think that for research purposes the existing MatLab based implementation is good enough to test most important characteristics of the developed optimization framework for hexapod. One should also mention that the task of infinity norm $\| \mathbf{J}^{-1} \|_{\infty}$ minimization could be also solved using cascaded loops of iteration with predefined steps (*e.g.*, $\Delta x = 0.001$ mm, $\Delta y = 0.001$ mm, $\Delta z = 0.001$ mm, $\Delta \alpha = 0.0001$ rad, $\Delta \beta = 0.0001$ rad, $\Delta \gamma = 0.0001$ rad) for all allowed end-effector coordinates x $[x_1 .. x_n]$, y $[y_1 .. y_n]$, z $[z_1 .. z_n]$, α $[\alpha_1 .. \alpha_n]$, β $[\beta_1 .. \beta_n]$ and γ $[\gamma_1 .. \gamma_n]$ within the hexapod workspace. Such scanning of the whole hexapod robot workspace would have its benefits (*e.g.*, it would provide the global view over the hexapod accuracy) and drawbacks (*e.g.*, limited flexibility, low accuracy, because if one takes very small iterations of end-effector coordinates, then one will not be able to accomplish such task with a normal PC, *etc.*). We decided not to implement this approach of the global scanning of the hexapod workspace, because we presented a similar simulation scheme with the iteration of joint coordinates as applied to PUMA-560 robotic manipulator (see Chapter 3.3 for details).

4.4. COMPUTER SIMULATION OF HEXAPOD ROBOT POSITIONING ACCURACY IMPROVEMENT USING JOINT ERROR MAXIMUM MUTUAL COMPENSATION

If hexapod Jacobian $\mathbf{J}$ is known, then the infinity norm $\| \mathbf{J}^{-1} \|_{\infty}$ can be easily obtained as explained further. Jacobian $\mathbf{J}$ for hexapod robot is a 6 x 6 matrix (Angeles, 2003):

$$\mathbf{J}(\mathbf{x}) = \begin{bmatrix} j_{11} & j_{12} & j_{13} & j_{14} & j_{15} & j_{16} \\ j_{21} & j_{22} & j_{23} & j_{24} & j_{25} & j_{26} \\ j_{31} & j_{32} & j_{33} & j_{34} & j_{35} & j_{36} \\ j_{41} & j_{42} & j_{43} & j_{44} & j_{45} & j_{46} \\ j_{51} & j_{52} & j_{53} & j_{54} & j_{55} & j_{56} \\ j_{61} & j_{62} & j_{63} & j_{64} & j_{65} & j_{66} \end{bmatrix}, \tag{4.16}$$

where **x** $(x, y, z, \alpha, \beta, \gamma)$ is the vector of Cartesian positions and orientation angles (roll, pitch and yaw) of hexapod end-effector. The elements j_{ij} of hexapod Jacobian were calculated using Symbolic Math Toolbox from MatLab (see Appendix G for details). The detailed expressions for j_{ij} can be found in Appendix A.

The inverse of hexapod Jacobian $\mathbf{J}^{-1}$ can be presented as:

$$\mathbf{J}^{-1}(j_{ij}) = \begin{bmatrix} a_{11} & a_{12} & a_{13} & a_{14} & a_{15} & a_{16} \\ a_{21} & a_{22} & a_{23} & a_{24} & a_{25} & a_{26} \\ a_{31} & a_{32} & a_{33} & a_{34} & a_{35} & a_{36} \\ a_{41} & a_{42} & a_{43} & a_{44} & a_{45} & a_{46} \\ a_{51} & a_{52} & a_{53} & a_{54} & a_{55} & a_{56} \\ a_{61} & a_{62} & a_{63} & a_{64} & a_{65} & a_{66} \end{bmatrix}, \tag{4.17}$$

where i and j are indices [1 .. 6] showing that each of a_{ij} in (4.21) is dependent on j_{ij}, because $\mathbf{J}^{-1}$ is the inverse of **J**. We do not show the details of elements a_{ij} because expressions formed after inversion of (4.17) would require additional 120 pages. We simply refer those interested to obtain the detailed expressions of Jacobian inverse to MatLab Symbolic Math Toolbox. One should define j_{ij} as symbols and then use *inv*() and *ccode*() functions of MatLab to get the detailed expressions of Jacobian inverse.

By definition, the infinity norm $\| \mathbf{J}^{-1} \|_{\infty}$ is (Chong and Zak, 2001):

$$\| \mathbf{J}^{-1} \|_{\infty} = \max_{i=1..6} \sum_{k=1}^{6} |a_{ik}| , \tag{4.18}$$

If we define sums A_i for each of matrix rows as:

$$A_i = \sum_{k=1}^{6} |a_{ik}| , \; i = 1 .. 6$$

then

$$\| \mathbf{J}^{-1} \|_{\infty} = \max (A_1, A_2, A_3, A_4, A_5, A_6) . \tag{4.19}$$

Based on (4.16) and (4.17), it is clear that $A_i = f_i(x, y, z, \alpha, \beta, \gamma)$. At this point, we are able to present the whole Algorithm 4.2 used to calculate the minimum of infinity norm $\| \mathbf{J}^{-1} \|_{\infty}$ in MatLab (see Appendix A for details of implementation in MatLab). Algorithm 4.2 is largely based on Algorithm 4.1.

Algorithm 4.2

1. Determine the working point $\mathbf{P}_{initial}$ $(x, y, z, \alpha, \beta, \gamma)$ for hexapod end-effector in the given part of the robot workspace (usually, the choice of this point is limited by the location of technological equipment).
2. Calculate Jacobian matrix **J** of hexapod for $\mathbf{P}_{initial}$ $(x, y, z, \alpha, \beta, \gamma)$.
3. Calculate sums for each of rows A_i ($i = 1 .. 6$) based on Jacobian matrix **J** values.
4. Identify maximum among A_i.
5. Use the maximum $A_i = f_i(x, y, z, \alpha, \beta, \gamma)$ found on step 4 as the infinity norm $\|\mathbf{J}^{-1}\|_\infty$, and minimize its $f_i(x, y, z, \alpha, \beta, \gamma)$ using BFGS algorithm.
   a. On each step of BFGS algorithm check if A_i is still an infinity norm $\| \mathbf{J}^{-1} \|_\infty$. If yes, then continue minimization. If no, then remember current coordinates $(x, y, z, \alpha, \beta, \gamma)$ as critical ones (these coordinates can be later used as intermediate "success", if the minimization process does not converge further for any reasons) and go to step 4 to continue minimization with another A_i which became the infinity norm $\| \mathbf{J}^{-1} \|_\infty$. One should mention that MatLab internally implements this logic in its Optimization Toolbox.
   b. Continue minimization until the local minimum for infinity norm $\| \mathbf{J}^{-1} \|_\infty$ is found with the required level of accuracy. This means the coordinates of the new point $\mathbf{P}_{final}$ $(x', y', z', \alpha', \beta', \gamma')$ will become available.
6. Under the given exemplary joint errors $\Delta\mathbf{q}$, one can compare the difference between end-effector positioning and orientation accuracy (ISO 9283, 1998) in the initial point $\mathbf{P}_{initial}$ $(x, y, z, \alpha, \beta, \gamma)$ and final point $\mathbf{P}_{final}$ $(x', y', z', \alpha', \beta', \gamma')$ to quantify end-effector accuracy improvement in the new point.

It is important to mention that Algorithm 4.2 is suitable for so-called unconstrained hexapod robot in which we do not set any limitations on joint coordinates and end-effector orientation. In sub-chapter 4.5 we will present an example of the constrained hexapod robot more relevant to the practice.

In this research, we performed a number of simulations on hexapod model in MatLab to identify possible improvements of end-effector positioning and orientation accuracy in $\mathbf{P}_{final}$ found based on Algorithm 4.2. The most important MatLab program listings for unconstrained hexapod robot can be found in Appendix A. We defined parameters of our hexapod robot used in simulations by setting radii of the hexapod base and end-effector to $r_b = 0.076$ m and $r_e = 0.03$ m, accordingly, height of the platform in neutral position to $h = 0.26$ m and angles between the first two joints on the base and end-effector to $\theta_b = 30°$ and $\theta_e = -60°$, accordingly (see Fig 4.6 for details).

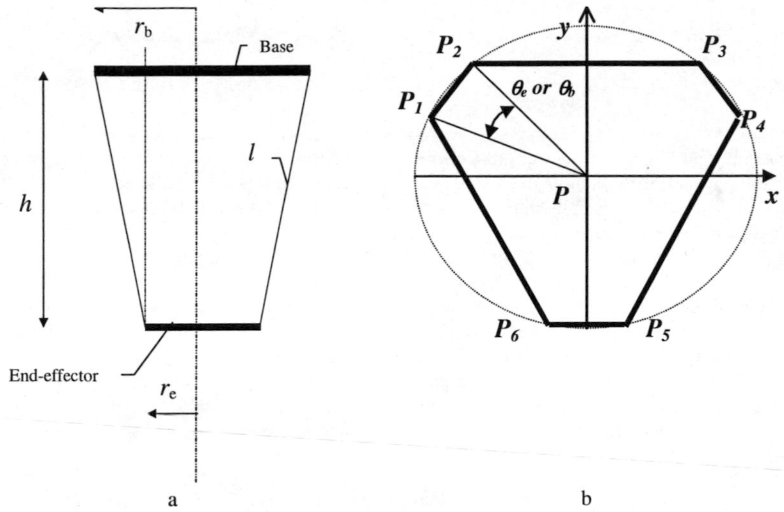

Fig. 4.6. Geometry of unconstrained hexapod robot used in simulations:
a – height h in neutral position and radii r_b and r_e of hexapod base and end-effector (side view),
b – angles θ_b and θ_e between the first two joints on the base and end-effector (front view).

The lengths $l \sim 0.28$ m of hexapod legs in neutral position can be easily found using above-mentioned hexapod parameters and simple geometric operations with triangles formed by hexapod base, end-effector and legs. Such a hexapod robot has a hexagon as its base and end-effector. We defined four initial hexapod end-effector poses (freely and randomly selected in the hexapod robot workspace) with coordinates $\mathbf{P}_{\text{initial}}$ $(x, y, z, \alpha, \beta, \gamma)$ for our unconstrained hexapod robot simulations:

$\mathbf{P}_{\text{initial}}^1$ (-0.02 m, -0.01 m, 0.26 m, 0.1 rad, 0.05 rad, -0.07 rad), **Infinity norm**: 166.98

$\mathbf{P}_{\text{initial}}^2$ (0.05 m, 0.03 m, 0.18 m, 0.3 rad, -0.1 rad, 0.17 rad), **Infinity norm**: 129.91

$\mathbf{P}_{\text{initial}}^3$ (0.12 m, -0.13 m, 0.14 m, 0.02 rad, -0.23 rad, -0.29 rad), **Infinity norm**: 192.35

$\mathbf{P}_{\text{initial}}^4$ (0.08 m, 0.06 m, 0.22 m, -0.4 rad, -0.25 rad, -0.12 rad), **Infinity norm**: 167.48

in which the minimization of the infinity norm of hexapod Jacobian inverse was initialized. For our simulations we used IBM Laptop with 2MB RAM, 20GB free hard disk space and ~1.6 GHz CPU. It took ~ 30 s to get results of minimization from MatLab application. We obtained the following values of $\mathbf{P}_{\text{final}}$ $(x', y', z', \alpha', \beta', \gamma')$:

$\mathbf{P}_{\text{final}}^1$ (-0.006 m, -0.005 m, 0.076 m, 0.097 rad, 0.05 rad, -0.069 rad), **Infinity norm**: 62.98

$\mathbf{P}_{\text{final}}^2$ (-0.018 m, -0.036 m, 0.065m, -0.009 rad, -0.065 rad, 0.317 rad), **Infinity norm**: 57.73

$\mathbf{P}_{final}^3$ (-0.028 m, 0.031 m, 0.061 m, 0.044 rad, -0.155 rad, -0.253 rad), **Infinity norm**: 57.74

$\mathbf{P}_{final}^4$ (-0.008 m, -0.057 m, -0.056 m, 0.025 rad, 0.032 rad, -0.501 rad), **Infinity norm**: 55.96.

More details about the computation process of infinity norm minimization for hexapod Jacobian inverse are shown in graphs in Fig. 4.7. As one can see from the above-presented results, the local minimums of infinity norm of hexapod Jacobian inverse were successfully found. The value of infinity norm was significantly reduced, *e.g.*, by 63 % in $\mathbf{P}_{final}^1$, by 56 % in $\mathbf{P}_{final}^2$, by 70 % in $\mathbf{P}_{final}^3$ and by 67 % in $\mathbf{P}_{final}^4$. One should pay special attention to *Z* coordinate of hexapod end-effector. The value of this coordinate significantly decreases in all $\mathbf{P}_{final}$ poses comparing to $\mathbf{P}_{initial}$. One can interpret this as a fact that the best joint error mutual compensation takes place in hexapod robots when they are in a "sitting" position, *e.g.*, when *Z* coordinate of hexapod end-effector is small. This may be a limitation in a number of hexapod applications where the whole hexapod robot workspace should be used.

To verify that positioning and orientation accuracies in poses $\mathbf{P}_{final}$ are better than those in $\mathbf{P}_{initial}$, we will find positioning and orientation accuracies (ISO 9283) in all four $\mathbf{P}_{initial}$ and $\mathbf{P}_{final}$ and compare those values. We will use ten random combinations of joint errors $\Delta\mathbf{q}$ (in meters) for this purpose:

$\Delta\mathbf{q}_1$ (0.0001; -0.0005; 0.0005; -0.0004; 0.0002; 0.0003),

$\Delta\mathbf{q}_2$ (-0.0008; -0.0001; 0.00004; -0.0001; -0.0007; -0.0006),

$\Delta\mathbf{q}_3$ (-0.0001; -0.0007; -0.00005; -0.00004; -0.0009; -0.0003),

$\Delta\mathbf{q}_4$ (0.0001; 0.0005; 0.0005; 0.0009; 0.0008; 0.0009),

$\Delta\mathbf{q}_5$ (0.0001; 0.0001; 0.0001; 0.0001; 0.0001; 0.0001),

$\Delta\mathbf{q}_6$ (-0.0002; -0.0002; -0.0001; -0.0001; -0.0001; -0.0001),

$\Delta\mathbf{q}_7$ (0.0002; -0.0002; 0.0004; 0.0009; -0.00002; 0.000003),

$\Delta\mathbf{q}_8$ (-0.0005; -0.0005; -0.0005; 0.0004; 0.0004; 0.0004),

$\Delta\mathbf{q}_9$ (0.0000001; -0.000005; 0.000002; -0.0000004; 0.000001; 0.0008) and

$\Delta\mathbf{q}_{10}$ (-0.00001; -0.0000005; -0.00005; -0.00002; 0.0002; 0.00002).

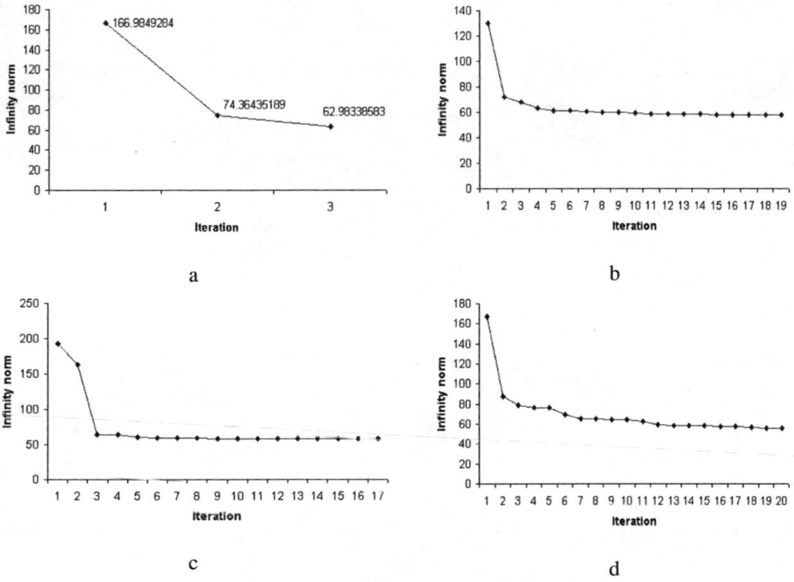

Fig. 4.7. Minimization of infinity norm of hexapod Jacobian inverse using BFGS algorithm:
a – in $\mathbf{P}_{initial}^{1}$ (after 3 iterations and 231 function evaluations),
b – in $\mathbf{P}_{initial}^{2}$ (after 19 iterations and 469 function evaluations),
c – in $\mathbf{P}_{initial}^{3}$ (after 17 iterations and 686 function evaluations) and
d – in $\mathbf{P}_{initial}^{4}$ (after 20 iterations and 623 function evaluations).

Comparison results are presented in Tables 4.2 – 4.5. As one can see from the results in Tables 4.2 – 4.5, end-effector positioning and orientation accuracy was significantly improved in $\mathbf{P}_{final}$ in most cases (*e.g.*, positioning accuracy was improved by ~ 40 % and orientation accuracy was improved by ~ 20 % in average), because of improved joint error mutual compensation in poses with infinity norm minimum of hexapod Jacobian inverse. However, for some combinations of joint errors (see grey cells in Tables 4.2 – 4.5) the deterioration of end-effector positioning and/or orientation accuracy was observed. Such accuracy deterioration was expected for a minor number of joint error combinations similar to the case with the 2-R robotic manipulator (see Table 2.3). Another very important aspect, which can be observed from results presented in Tables 4.2 – 4.5, is that the deterioration of end-effector accuracy takes place when joint errors are very small. On the opposite, the best improvement of end-effector accuracy takes place when the initial joint errors are big. This is exactly as we expected, because the presented minimization of Jacobian infinity norm was designed for worst case joint errors.

Table 4.2

Comparison of positioning and orientation accuracy in $\mathbf{P}_{initial}^{1}$ and $\mathbf{P}_{final}^{1}$

Joint errors	Initial positioning accuracy [m] x 10^{-6}	Final positioning accuracy [m] x 10^{-6}	Positioning improvement / deterioration [%]	Initial orientation accuracy [rad] x 10^{-3}	Final orientation accuracy [rad] x 10^{-3}	Orientation improvement / deterioration [%]
$\Delta \mathbf{q}_1$	883.2	326.1	63.1	41.6	22.2	46.6
$\Delta \mathbf{q}_2$	1954.7	827.0	57.7	19.1	8.2	56.9
$\Delta \mathbf{q}_3$	858.5	509.0	40.7	17.5	21.9	-25.3
$\Delta \mathbf{q}_4$	1334.5	894.5	32.9	25.5	10.7	57.9
$\Delta \mathbf{q}_5$	102.4	126.1	-23.1	0.0387	0.116	-199.8
$\Delta \mathbf{q}_6$	215.0	178.1	17.15	1.46	1.67	-14.8
$\Delta \mathbf{q}_7$	1155.0	483.4	58.1	15.9	19.7	-24.0
$\Delta \mathbf{q}_8$	2605.5	949.5	63.5	25.0	9.2	63.1
$\Delta \mathbf{q}_9$	1337.4	514.9	61.5	24.3	14.7	39.4
$\Delta \mathbf{q}_{10}$	350.2	134.1	61.7	4.8	3.9	18.7
Σ/10			**43.3**			**1.8**

Table 4.3

Comparison of positioning and orientation accuracy in $\mathbf{P}_{initial}^{2}$ and $\mathbf{P}_{final}^{2}$

Joint errors	Initial positioning accuracy [m] x 10^{-6}	Final positioning accuracy [m] x 10^{-6}	Positioning improvement / deterioration [%]	Initial orientation accuracy [rad] x 10^{-3}	Final orientation accuracy [rad] x 10^{-3}	Orientation improvement / deterioration [%]
$\Delta \mathbf{q}_1$	623.3	280.1	55.1	38.3	21.5	43.8
$\Delta \mathbf{q}_2$	1525.1	755.0	50.5	15.0	8.4	44.2
$\Delta \mathbf{q}_3$	747.9	494.1	33.9	18.7	21.8	-16.8
$\Delta \mathbf{q}_4$	1130.4	862.7	23.7	19.6	10.7	45.1
$\Delta \mathbf{q}_5$	104.3	127.7	-22.4	0.17	0.14	18.1
$\Delta \mathbf{q}_6$	190.0	192.6	-1.4	1.74	1.80	-3.2
$\Delta \mathbf{q}_7$	856.3	531.3	37.9	16.9	18.7	-10.5
$\Delta \mathbf{q}_8$	1897.0	1040.0	45.2	21.5	11.0	48.7
$\Delta \mathbf{q}_9$	1024.2	522.0	49.0	21.2	16.8	20.8
$\Delta \mathbf{q}_{10}$	248.7	135.9	45.3	4.2	4.4	-3.1
Σ/10			**31.7**			**18.7**

Table 4.4

Comparison of positioning and orientation accuracy in $\mathbf{P}_{initial}{}^{3}$ and $\mathbf{P}_{final}{}^{3}$

Joint errors	Initial positioning accuracy [m] x 10^{-6}	Final positioning accuracy [m] x 10^{-6}	Positioning improvement / deterioration [%]	Initial orientation accuracy [rad] x 10^{-3}	Final orientation accuracy [rad] x 10^{-3}	Orientation improvement / deterioration [%]
Δq_1	1460.2	346.5	76.2	79.6	20.0	74.8
Δq_2	2562.7	964.6	62.3	31.2	11.5	63.2
Δq_3	1058.5	513.8	51.5	53.3	22.2	58.3
Δq_4	1400.0	942.7	32.7	44.0	12.7	71.0
Δq_5	102.4	129.4	-26.32	0.0823	0.15	-88.3
Δq_6	206.7	177.1	14.3	2.9	2.0	30.1
Δq_7	812.8	479.7	41.0	53.2	21.3	60.0
Δq_8	2719.1	1034.8	62.0	41.0	10.6	74.1
Δq_9	1510.1	476.7	68.4	30.1	13.3	55.8
Δq_{10}	396.1	130.7	67.1	9.1	3.6	60.2
Σ/10			**44.9**			**45.9**

Table 4.5

Comparison of positioning and orientation accuracies in $\mathbf{P}_{initial}{}^{4}$ and $\mathbf{P}_{final}{}^{4}$

Joint errors	Initial positioning accuracy [m] x 10^{-6}	Final positioning accuracy [m] x 10^{-6}	Positioning improvement / deterioration [%]	Initial orientation accuracy [rad] x 10^{-3}	Final orientation accuracy [rad] x 10^{-3}	Orientation improvement / deterioration [%]
Δq_1	926.7	248.2	73.2	42.7	22.5	47.4
Δq_2	1912.0	738.2	61.4	21.1	7.4	64.9
Δq_3	963.6	481.8	50.0	20.8	20.5	1.2
Δq_4	1206.4	849.1	29.6	29.5	10.7	63.7
Δq_5	102.4	126.1	-23.1	0.19	0.36	-88.0
Δq_6	211.7	199.5	5.7	1.7	1.9	-10.6
Δq_7	1058.6	567.2	46.4	15.4	17.6	-13.7
Δq_8	2347.1	1145.3	51.2	27.3	14.5	47.0
Δq_9	1413	549.5	61.1	29.2	17.4	40.4
Δq_{10}	352.4	159.1	54.9	6.5	4.4	31.9
Σ/10			**41.0**			**18.4**

4.5. COMPUTER SIMULATION OF POSITIONING ACCURACY IMPROVEMENT FOR REAL HEXAPOD ROBOT

In this sub-chapter, we present similar simulation results to those in sub-chapter 4.4, but this time for a constrained hexapod robot, which is very close to practice. We will use the hexapod model in MatLab to identify possible improvements of end-effector positioning and orientation accuracy in $\mathbf{P}_{final}$ found based on the modified Algorithm 4.2. The main modifications in Algorithm 4.2 are:

- On each iteration of the BFGS algorithm we will calculate lengths of hexapod legs and make sure that they are within the allowed range;
- We will fix the orientation of hexapod end-effector.

We defined parameters of our hexapod robot used in simulations by setting radii of the hexapod base and end-effector to r_b = 0.076 m and r_e = 0.03 m, accordingly, height of the platform in neutral position to h = 0.26 m and angles between the first two joints on the base and end-effector to $\theta_b = 30°$ and $\theta_e = 0°$, accordingly (see Fig 4.6 for details). The lengths of hexapod legs in the neutral position were easily found to be $l \sim 0.265$ m. The allowed range of hexapod leg lengths is [0.04 .. 0.3] m. Such a hexapod robot has the hexagon as the base and triangular end-effector (see Fig. 4.4).

We defined one initial hexapod end-effector pose (freely and randomly selected in the hexapod robot workspace) with coordinates $\mathbf{P}_{initial}$ $(x, y, z, \alpha, \beta, \gamma)$ for our constrained hexapod robot simulation:

$\mathbf{P}_{initial}$ (-0.02 m, -0.01 m, 0.26 m, 0.174 rad, 0 rad, 0 rad), **Infinity norm**: 456.62

in which the minimization of the infinity norm of hexapod Jacobian inverse was initialized. We obtained the following values of $\mathbf{P}_{final}$ $(x', y', z', \alpha', \beta', \gamma')$:

$\mathbf{P}_{final}$ (-0.0049 m, -0.0024 m, 0.061 m, 0.174 rad, 0 rad, 0 rad), **Infinity norm**: 137.03

The initial and final positions of the hexapod robot are shown in Fig. 4.8.

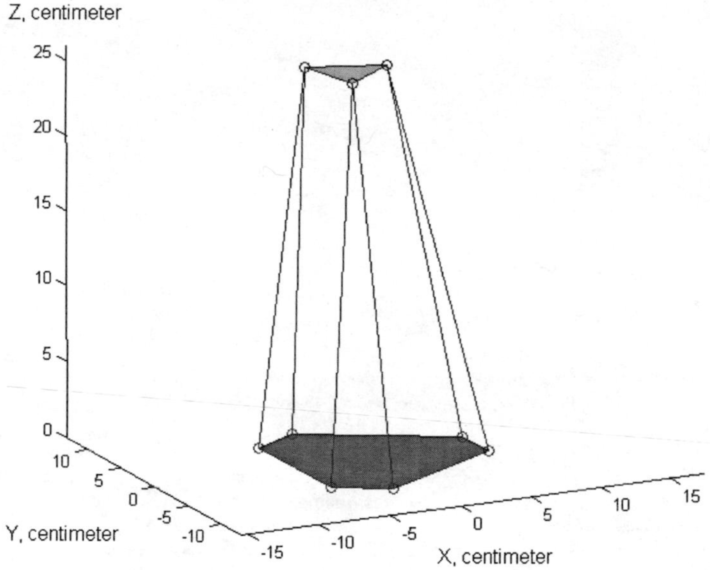

a

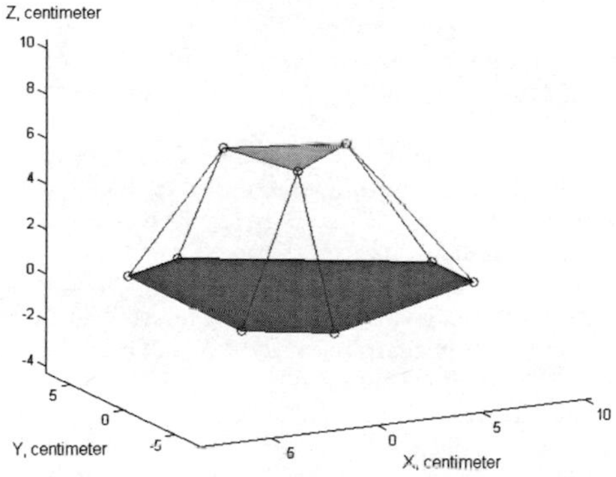

b

Fig.4.8. Poses of constrained hexapod robot: a – initial pose $\mathbf{P}_{initial}$ and b – final pose $\mathbf{P}_{final}$

The computation process of infinity norm minimization for hexapod Jacobian inverse is shown in the graph in Fig. 4.9.

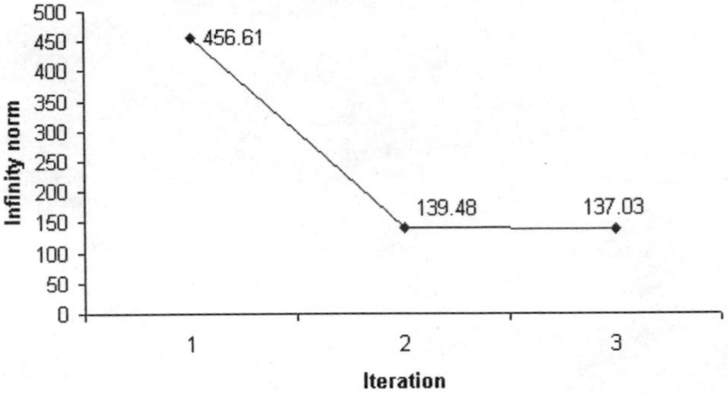

Fig. 4.9. Minimization of infinity norm of hexapod Jacobian inverse using BFGS algorithm in $\mathbf{P}_{initial}$ (after 3 iterations and 148 function evaluations)

In our example with l [0.04 .. 0.3] m, the local minimum of infinity norm of hexapod Jacobian inverse was successfully found. The value of the infinity norm was significantly reduced by 70 % in $\mathbf{P}_{final}$. To verify that positioning and orientation accuracy in pose $\mathbf{P}_{final}$ is better than that in $\mathbf{P}_{initial}$, we will find positioning and orientation accuracy (ISO 9283) in $\mathbf{P}_{initial}$ and $\mathbf{P}_{final}$ and compare them. We will use the same ten random combinations of joint errors $\Delta\mathbf{q}$ (in meters) as those in sub-chapter 4.4. Comparison results are presented in Table 4.6. As one can see from the results in Table 4.6, end-effector positioning and orientation accuracy was significantly improved in $\mathbf{P}_{final}$ in most cases (*e.g.*, positioning accuracy was improved by ~ 45 % and orientation accuracy was improved by ~ 3 % in average), because of improved joint error mutual compensation in poses with smaller infinity norm of hexapod Jacobian inverse. However, for some combinations of joint errors (see grey cells in Table 4.6) deterioration of end-effector positioning and/or orientation accuracy was observed, as expected.

As one can see in Fig. 4.9, the first jump in the BFGS algorithm is quite big and it is possible that in real hexapod models the allowed range of leg length will be exceeded already on the first step of BFGS algorithm. Thus, it is highly recommended to use one of the constrained optimization algorithms for real hexapod robots (Chong and Zak, 2001) in practice. As one could see from results of $\|\mathbf{J}^{-1}\|_\infty$ minimization in Tables 4.2 – 4.5, Z coordinate of hexapod end-effector contributed most significantly to the minimization.

The obtained results confirm that joint error maximum mutual compensation can be quite effective if one aims to improve end-effector positioning accuracy. The same approach with minimization of infinity norm of manipulator Jacobian inverse or Jacobian (depending on manipulator type) can be applied for other types of robotic manipulators. The changing values of joint errors remain the major open issue. In practice, it is still possible that in robot configurations with minimum infinity norm of Jacobian or Jacobian inverse (depending on manipulator type) joint errors may significantly increase. This may deteriorate the effect of joint error maximum mutual compensation. It will also significantly complicate any experimental measurements to verify joint error maximum mutual compensation in practice.

Table 4.6

Comparison of positioning and orientation accuracy in $\mathbf{P}_{initial}$ and $\mathbf{P}_{final}$ for constrained hexapod robot

Joint errors	Initial positioning accuracy [m] x 10^{-6}	Final positioning accuracy [m] x 10^{-6}	Positioning improvement / deterioration [%]	Initial orientation accuracy [rad] x 10^{-3}	Final orientation accuracy [rad] x 10^{-3}	Orientation improvement / deterioration [%]
$\Delta\mathbf{q}_1$	4014.00	1178.85	70.6	109.91	48.19	56.1
$\Delta\mathbf{q}_2$	3342.45	1040.24	68.9	51.67	19.52	62.2
$\Delta\mathbf{q}_3$	4688.48	1427.63	69.5	36.68	46.06	-25.6
$\Delta\mathbf{q}_4$	1539.93	982.20	36.2	68.85	20.76	69.8
$\Delta\mathbf{q}_5$	101.60	126.00	-24.0	0.066	0.22	-241.5
$\Delta\mathbf{q}_6$	135.14	166.92	-23.5	2.24	2.76	-23.2
$\Delta\mathbf{q}_7$	3526.64	1065.32	69.8	32.11	38.80	-20.8
$\Delta\mathbf{q}_8$	4114.08	1244.25	69.7	68.67	21.42	68.8
$\Delta\mathbf{q}_9$	3624.48	1088.85	69.9	65.61	34.01	48.2
$\Delta\mathbf{q}_{10}$	913.15	276.87	69.7	12.34	8.34	32.4
Σ/10			**47.68**			**2.64**

The presented approach can find its practical implementation in robotic systems with hexapods and other similar robotic manipulators. In particular, the joint error maximum mutual compensation can be widely used on the stages of robot end-effector trajectory planning and definition of robot workspace.

5. TYPICAL ROBOTIC SYSTEMS FOR APPLICATION OF JOINT ERROR MAXIMUM MUTUAL COMPENSATION

The developed method of the end-effector positioning accuracy improvement using joint error maximum mutual compensation is primarily aimed to be used in the robotic manipulators with open-loop control systems and stepping drives in joints (*e.g.*, in SCARA type robots (see Fig. 0.3)), because in such control systems there are no feedback sensors and, as it is for example in stepping drives, such control systems rely on the precision of the execution of control pulses coming from the control system to the rotor or stator of the motor. However, the developed method is still applicable for any other types (*e.g.*, AC or DC) of drives used in robotic manipulator joints.

An exemplary application area for joint error maximum mutual compensation is the hexapod (see also Chapter 1.2 for more details), *e.g.*, from the Microbotic platform (see Fig. 5.1), which was the part of the industrial Ph.D. dissertation done by Kasper Doering (Doering, 2004) at the University of Southern Denmark. Microbotic platform is a fully automated platform for micro-handling and processing of micro-parts, for example, in hearing aid assembly.

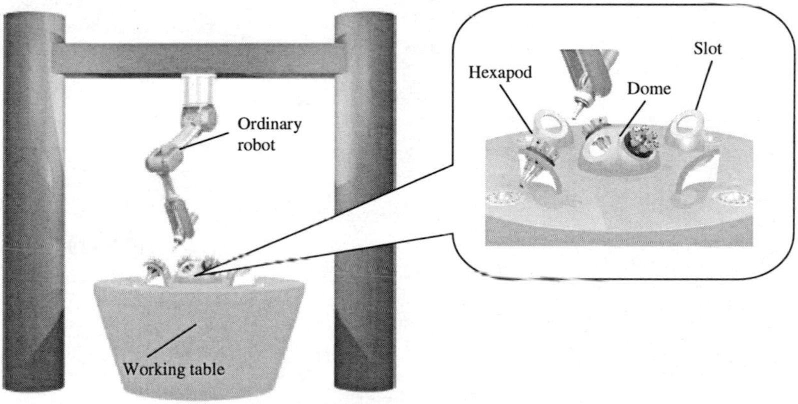

Fig. 5.1. Multiple hexapod robots brought together to solve complex micro-handling task

Slots (see Fig. 5.1) provide mounting points for hexapods, and only when mounted in a slot, a hexapod will be able to move. The hexapods are mobile in the sense that it is possible to move them between different slots while the workcell is working. By using a miniature version of the hexapod (see Fig. 1.4), we get the necessary dexterity to perform arbitrary motions in 3-D space and the rigidity to obtain the accuracy that we need. An ordinary industrial robot hanging over the workcell suspended by an iron portal takes care of hexapod translocations. Inside the dome it is possible to control environmental parameters such as light, temperature, dust and atmosphere, which quite often constitute important factors in micro-handling and processing. Complex micro-handling can, thus, be done in the shelter of the dome by bringing together multiple hexapod robots for co-operation (Doering, 2004).

Obtaining the necessary accuracy is a difficult task requiring not only development of highly specialized and accurate mechanics but also the advanced control. The control of the hexapod relies on the feedback provided by an unusual combination of sensor systems. In short, two different types of sensors are used. An external optical sensor system that measures the pose of the tool of the hexapod relative to some global workcell reference. In addition, a little video camera is mounted near the tool of the hexapod, as shown in Fig. 1.4. When the hexapod approaches the expected position of a certain workpiece, the object enters the field of camera view. Using advanced computer vision algorithms the video stream is analyzed and the pose of the tool relative to relevant features of the workpiece is calculated (Doering, 2004). These two measurements work in conjunction to provide highly accurate probing of the pose of the hexapod tool in order to enable relevant control.

Our proposal for such applications like Microbotic platform is to use joint error maximum mutual compensation for both the serial robot and hexapod. This means the technological operations will have to be executed by the ordinary robot (see Fig. 5.1) and hexapod in such configurations (or in close neighborhoods of those configurations) in which the best joint error mutual compensation takes place. This will provide higher accuracy (the estimated improvement is 15 – 20 %) of the technological operation execution by Microbotic platform.

CONCLUSIONS

The method of end-effector positioning accuracy improvement using joint error maximum mutual compensation was developed for robotic manipulators. The use of joint error maximum mutual compensation allowed improving robot end-effector positioning accuracy by 1 to 15 percent in the given area of the robot workspace and can be implemented when joint errors of robotic manipulator are known or unknown. Appropriate computer simulations and experiments have been performed for 2-R SCARA type robotic manipulator to confirm the theory. The core of the developed method is to take into account joint error maximum mutual compensation when one determines robot end-effector working points in the robot workspace. This includes two cases. First, one can use joint error maximum mutual compensation to compensate some known or predictable joint errors (when it is impossible or inconvenient to compensate them at the control system level or trajectory planning stage). Second, one can use joint error maximum mutual compensation to improve end-effector positioning accuracy by compensating unknown or only partly known joint errors. If joint error values are not known or only partly known then, with the probability of 0.75, the usage of joint error maximum mutual compensation for 2-R robotic manipulators will improve end-effector positioning accuracy. The latter case with unknown joint errors is universal and is valid for all kinds of robotic manipulators that have at least two rotational joints in one plane. The theoretical maximum of end-effector positioning accuracy improvement for a 2-R SCARA type robotic manipulator using joint error maximum mutual compensation is quite high (for example, positioning accuracy improvement could be up to 90.4 % for a 2-R SCARA type robotic manipulator), if one could change the position of robot end-effector within the whole robot workspace and joint errors would remain unchanged (in practice, such case is not common (Kieffer, Cahill and James, 1997) because joint errors may significantly change with the change of the end-effector position in the robot workspace and the optimal coordinate $q_{2opt} = 180°$ is on the border of the robot workspace). The implementation of joint error maximum mutual compensation is the most effective for 2-R SCARA type robotic manipulators and not for complex robotic manipulators that include 2-R chains, because with the increase of the number of 2-R chains the probability of end-effector positioning accuracy improvement in all 2-R chains decreases significantly and already for robotic manipulator with three 2-R chains it is only 0.42 (lower than 0.5).

We presented the worst case error approach for joint error maximum mutual compensation and developed a mathematical model that allowed predicting usability of joint error maximum mutual compensation with worst case error approach for any 2-R planar robotic manipulator. Based on the example with industrial SCARA type robot RM 10-01, we showed that joint error maximum mutual compensation with worst case error approach would provide significant improvement (20 – 50 %) of end-effector positioning accuracy if joint error Δq_2 would not change by more than 50 – 80 % in the new point (the exact values depend on initial conditions).

We investigated the application of joint error maximum mutual compensation for hexapod robots. We developed the algorithm and simulation framework in MatLab to find the optimal hexapod configurations with joint error maximum mutual compensation in which the positioning accuracy of hexapod end-effector is the best. The optimal hexapod configurations were found using the local minimum of the infinity norm of hexapod Jacobian inverse. The usage of joint error maximum mutual compensation in hexapods can improve hexapod end-effector positioning accuracy by ~ 40 % in average. The same approach with the minimization of infinity norm of manipulator Jacobian inverse or Jacobian (depending on manipulator type) can be applied for other types of robotic manipulators. The changing values of joint errors remain the major open issue. In practice, it is still possible that in robot configurations with minimum infinity norm of Jacobian or Jacobian inverse (depending on manipulator type) joint errors may significantly increase. This may deteriorate the effect of joint error maximum mutual compensation. It will also significantly complicate any experimental measurements to verify joint error maximum mutual compensation in practice. The presented approach can find its practical implementation in robotic systems with hexapods and other similar robotic manipulators. In particular, the joint error maximum mutual compensation can be widely used on the stages of robot end-effector trajectory planning and definition of robot workspace.

Advanced computer simulations were performed for hexapod robot and 6-DOF PUMA-560 type robotic manipulator to investigate the opportunities for a practical implementation of joint error maximum mutual compensation. Appropriate software tools, simulation framework and algorithms were developed and applied to test the benefits of joint error maximum mutual compensation. The simulation results showed that the positive effect of joint error maximum mutual compensation is dependent on joint error values and signs, and is only slightly dependent on the position of the command point in the robot workspace for 6-DOF PUMA-560 type robotic manipulator. Another important conclusion is that the benefit of application of joint error maximum mutual compensation for 6-DOF PUMA-560 type robotic manipulator is rather small. The end-effector positioning accuracy was improved by 3.5 % in the best case. The largest contribution to end-effector positioning accuracy improvement was observed from joint coordinates q_4 and q_5.

The practical areas and typical robotic systems where joint error maximum compensation could be applied were presented and investigated. The end-effector positioning accuracy improvement using joint error maximum compensation can be used for both robotic manipulators with closed-loop control systems and for robotic manipulators with open-loop control systems. The best effect of using joint error maximum mutual compensation can be achieved in robotic manipulators with open-loop control systems with stepping drives since they do not have feedback sensors and, as it is for example in stepping drives, rely on the precision of execution of control pulses coming from the control system to a rotor or stator of the stepping motor. Therefore, a common case for joint error maximum mutual compensation implementation is when stepping motors are used in SCARA robot joints. Typical technological operations for such SCARA robots are sealing, dispensing, parts insertion and assembly.

During our work on joint error maximum mutual compensation the following major issues were identified as subjects for future works and improvements:

- Integration of optimization algorithms for joint error maximum mutual compensation to robot OLP systems;
- Detailed industrial tests of the presented approach and its usability check in typical industrial applications;
- Evaluation of application and usability of joint error maximum mutual compensation not only for end-effector positioning, but also for end-effector orientation and robot path-tracking.

REFERENCES

1. Andreff N., Marchadier A. and Martinet P. Vision-based control of a Gough-Stewart parallel mechanism using legs observation // Proc. of IEEE Int. Conf. on Robotics and Automation. – Barcelona, Spain, 2005. – pp. 2546-2551.
2. Angeles J. Fundamentals of robotic mechanical systems. – New York, USA: Springer, 2002.
3. Baker D. and Wampler C. On the inverse kinematics of redundant manipulators // International Journal of Robotics Research – 1988. – Vol. 7, No. 9. – pp. 3-21.
4. Berg J. Path and orientation accuracy of industrial robots // International Journal of Advanced Manufacturing Technology. – 1993. – No. 8. – pp. 29-33.
5. Bernhardt R. and Albright S. Robot calibration. – London, UK: Chapman & Hall, 1993.
6. Besnard S. and Khalil W. Identifiable parameters for parallel robots kinematic calibration // Proc. of IEEE Int. Conf. on Robotics and Automation. – Seul, South Korea, 2001. – pp. 2859-2866.
7. Bobrow J. Optimal robot path planning using minimum time criterion // IEEE Trans. Rob. and Autom. – 1988. – Vol. 4, No. 7. – pp. 443-450.
8. Bobrow J., Dubowsky S. and Gibson J. Time–optimal control of robotic manipulators along specified paths // International Journal of Robotics Research. – 1985. – Vol. 4, No. 5. – pp. 3-17.
9. Bojadjiev T. Dynamic control of manipulating robots: Dynamic control by standard corrections // Prob. of Eng. Cybernetics and Robotics. – 1996. – No. 6. – pp. 87-97.
10. Borm J. and Menq C. Determination of optimal measurement configurations for robot calibration based on observability measure // International Journal of Robotics Research. – 1991. – Vol. 10, No. 1. – pp. 51-63.
11. Broderick P. and Cirpa R. A method for determining and correcting robot position and orientation errors due to manufacturing // Journal of Mechanisms, Transmissions and Automation in Design. – 1988. – Vol. 110. – pp. 3-10.
12. Chai K-S., Young K. and Tuersley I. A practical calibration process using partial information for a commercial Stewart platform // Robotica. – 2002. – Vol. 20, No. 3. – pp. 315-322.
13. Chevallereau C. Feasible trajectories in task space from a singularity for a non–redundant or redundant robot manipulator // International Journal of Robotics Research. – 1998. – Vol. 17, No. 1. – pp. 56-69.
14. Chong E. and Zak S. An introduction to optimization. – New York, USA: Wiley, 2001.
15. Dahl O. Path–constrained robot control with limited torque – Experimental evaluation // IEEE Trans. on Rob. and Autom. – 1994. – Vol. 10, No. 3. – pp. 658-669.
16. Dahl O. and Nielsen L. Torque–limited path following by online trajectory time scaling // IEEE Trans. Rob. and Autom. – 1990. – Vol. 6, No. 7. – pp. 554-561.
17. Daney D. Kinematic calibration of the Gough platform // Robotica. – 2003. – Vol. 21, No. 6. – pp. 677-690.
18. Daney D. and Emiris I. Robust parallel robot calibration with partial information // Proc. of IEEE Int. Conf. on Robotics and Automation. – Seul, South Korea, 2001. – pp. 3262-3267.
19. Daney D., Papegay Y. and Neumaier A. Interval methods for certification of the kinematic calibration of parallel robots // Proc. of IEEE Int. Conf. on Robotics and Automation. – New Orleans, USA, 2004. – pp. 1913-1918.
20. Diewald B., Godding R. and Henrich A. Robot calibration with a photogrammetric on-line system using reseau-scanning-cameras // In Optical 3-D Measurement Techniques II (eds. Grün/Kahmen). – Wichmann Verlag, Karlsruhe, Germany, 1993.

21. Dimov H., Dobrinov V. and Boiadjiev T. Experimental investigation of pose repeatability of manipulating robots // Prob. of Eng. Cybernetics and Rob. – 1997. – No. 6. – pp. 106-111.
22. Doering K. Control and coordination of robots in a flexible multi-robot platform with a special reference to micro-handling // Industrial Ph.D. dissertation. – University of Southern Denmark, Odense, Denmark, 2004.
23. Filonov I., Belajev G. and Veryha Y. A global approach to optimization in path–tracking for robotic manipulators // Proc. of Intelligent Manufacturing & Automation / DAAAM International. – Vienna, Austria, 1999. – pp. 157-158.
24. Filonov I., Kourtch L. and Veryha Y. Theoretical and experimental research of assembly center end-effector path-tracking // Proc. of 3d Int. Conf. Heavy machinery HM'99 / KUT. – Kraljevo, Yugoslavia, 1999. – pp. 312-315.
25. Filonov I. and Veryha Y. Numerical modelling of stress influence in rotational manipulator joints // Journal of Friction and Wear. – 2000. – Vol. 21, No. 6. – pp. 612-617. (in Russian)
26. Gmurman V. Theory of probabilities and mathematical statistics. – Moscow, Russia: High School, 2000. (in Russian)
27. Hayati S. and Mirmirani M. Improving the absolute positioning accuracy of robot manipulators // International Journal of Robotic Systems. – 1985. – No. 2. – pp. 397-413.
28. Hollerbach J. and Suh K. Redundancy resolution of manipulators through torque optimization // IEEE J. Rob. and Autom. – 1987. – Vol. RA–3, No. 4. – pp. 308-316.
29. Hollerbach J. and Wampler C. A taxonomy of kinematic calibration methods // International Journal of Robotics Research . – 1996. – Vol. 14. – pp. 573-591.
30. ISO 9283. Manipulating industrial robots – Performance criteria and related test methods. – IOS, Zurich, Switzerland, 1998.
31. Iurascu C. and Park F. Geometric algorithms for closed chain kinematic calibration // Proc. of IEEE Conf. on Robotics and Automation. – Detroit, USA, 1999. – pp. 1752-1757.
32. Jelenkovic L. and Budin L. Error analysis of a Stewart platform based manipulators // Proc. of 6th International Conference on Intelligent Engineering Systems. – Opatija, Croatia, 2002. – pp. 149-154.
33. Khalil W. and Besnard S. Self calibration of Stewart-Gough parallel robot without extra sensors // IEEE Trans. on Robotics and Automation. – 1999. – Vol. 15, No. 6. – pp. 1116-1121.
34. Kieffer J., Cahill A. and James M. Robust and accurate time–optimal path tracking control for robot manipulators // IEEE Trans. Rob. and Autom. – 1997. – Vol. 13, No. 6. – pp. 880-890.
35. Mirman C. and Gupta K. Identification of position independent robot parameter errors using special Jacobian matrices // International Journal of Robotics Research. – 1993. – Vol. 12, No. 3. – pp. 288-298.
36. Nahvi A., Hollerbach J. and Hayward V. Calibration of a parallel robot using multiple kinematics closed loops // Proc. of IEEE Int. Conf. on Robotics and Automation. – San Diego, USA, 1994. – pp. 407-412.
37. Nguyen C., Zhou Z. and Antrazi Z. Efficient Computation of Forward Kinematics and Jacobian Matrix of a Stewart Platform-Based Manipulator // Proc. of IEEE Southeast Conference. – Williamsburg, USA, 1991. – pp. 869-874.
38. Rauf A. and Ryu J. Fully autonomous calibration of parallel manipulators by imposing position constraint // Proc. of IEEE Int. Conf. on Robotics and Automation. – Seul, South Korea, 2001. – pp. 2389-2394.

39. Renaud P., Andreff N., Pierrot F. and Martinet P. Combining end-effector and legs observation for kinematic calibration of parallel mechanisms // Proc. of IEEE Int. Conf. on Robotics and Automation. – New Orleans, USA, 2004. – pp. 4116-4121.
40. Roth S., Mooring B. and Ravani B. An overview of robot calibration // IEEE Journal of Robotics and Automation. – 1987. – Vol. 3, No. 5. – pp. 377-385.
41. Shahinpur M. Course of robotics. – Moscow, Russia: Mir, 1990. (in Russian)
42. Slocum A. Precision machine design. – New Jersey, USA: Englewood Cliffs, 1992.
43. Smolnikov B. Problems of mechanics and optimization of robots. – Moscow, Russia: Science, 1990. (in Russian)
44. Tchon K. A normal form of singular kinematics of robot manipulators with smallest degeneracy // IEEE Trans. Rob. and Autom. – 1995. – Vol. 11, No. 3. – pp. 401-404.
45. Vaichav R. and Magrab E. A general procedure to evaluate robot positioning errors // International Journal of Robotics Research. – 1987. – Vol. 6, No. 1. – pp. 59-74.
46. Veryha Y. and Kurek J. Application of joint error mutual compensation for robot end-effector pose accuracy improvement // Journal of Intelligent and Robotic Systems. – 2003. – Vol. 36, No. 3. – pp. 315-329.
47. Veryha Y. and Kurek J. Robotic manipulator pose accuracy improvement using joint error maximum compensation // Proc. of IEEE Int. Conf. on Methods and Models in Automation and Robotics. – Szczecin, Poland, 2002. – pp. 993-998.
48. Vincze M., Filz K., Gander H., Prenninger J. and Zeichen G. A systematic approach to model arbitrary non-geometric kinematic errors // Advances in Robot Kinematics and Computational Geometry Series. – Kluwer Academic Publishers, Dordrecht, Netherlands, 1994. – pp. 129-138.
49. Vincze M., Prenninger J. and Gander H. A laser tracking system to measure position orientation of robot end effectors under motion // International Journal of Robotics Research. – 1994. – Vol. 13. – pp. 305-314.
50. Wehn H. and Belanger P. Ultrasound-based robot position estimation // IEEE Trans. Rob. and Autom. – 1997. – Vol. 13, No. 5. – pp. 682-692.
51. Yiu Y., Meng J. and Li Z. Auto-calibration for a parallel manipulator with sensor redundancy // Proc. of IEEE Int. Conf. on Robotics and Automation. – Taipei, Taiwan, 2003. – pp. 3660-3665.
52. Zhuang H. and Roth Z. A linear solution to the kinematic parameter identification of robot manipulators // IEEE Transactions in Robotics and Automation. – 1993. – Vol. 9, No. 2. – pp. 174-185.
53. Zhuang H., Yan J. and Masory O. Calibration of Stewart platform and other parallel manipulators by minimizing inverse kinematic residuals // Journal of Robotic Systems. – 1998. – Vol. 15, No. 7. – pp. 396-406.

APPENDIX A

Program Listings in MatLab

Elements of hexapod Jacobian matrix in MatLab syntax

We use the following symbols in expressions of Jacobian matrix elements:

- Rb, Rp - radii of the base and platform;
- Thetab, Thetap - angles between the first two joints on the base and platform;
- x; y; z; ax; ay; az – position and orientation coordinates (x, y, z, roll, pitch, yaw) of hexapod end-effector.

We present expressions only for elements $j_{11} - j_{16}$ of hexapod Jacobian. You can obtain the remaining expressions for other elements $j_{21} - j_{66}$ by contacting at yauheni.veryha@web.de.

j11

```
j11 = (-cos(ax)*cos(ay)*Rp*sin(0.3141592653589793E1/6.0+Thetap/2.0)-(-
sin(ax)*cos(az)+cos(ax)*sin(ay)*
sin(az))*Rp*cos(0.3141592653589793E1/6.0+Thetap/2.0)+x+Rb*sin(0.3141592653589793E1/6.0+T
hetab/2.0))/sqrt(power(-cos(ax)*cos(ay)*Rp*sin(0.3141592653589793E1/6.0+Thetap/2.0)-(-
sin(ax)*cos(az)+cos(ax)*sin(ay)*
sin(az))*Rp*cos(0.3141592653589793E1/6.0+Thetap/2.0)+x+Rb*sin(0.3141592653589793E1/6.0+T
hetab/2.0),2.0)+power(-sin(ax)*cos(ay)*Rp*sin(0.3141592653589793E1/6.0+Thetap/2.0)-
(cos(ax)*cos(az)+sin(ax)*sin(ay)*
sin(az))*Rp*cos(0.3141592653589793E1/6.0+Thetap/2.0)+y+Rb*cos(0.3141592653589793E1/6.0+T
hetab/2.0),2.0)+power(sin(ay)*Rp*sin(0.3141592653589793E1/6.0+Thetap/2.0)-cos(ay)*sin(az)*Rp*
cos(0.3141592653589793E1/6.0+Thetap/2.0)+z,2.0));
```

j12

```
j12 = (-sin(ax)*cos(ay)*Rp*sin(0.3141592653589793E1/6.0+Thetap/2.0)-
(cos(ax)*cos(az)+sin(ax)*sin(ay)*sin(az))*Rp*cos(0.3141592653589793E1/6.0+Thetap/2.0)+y+Rb*c
os(0.3141592653589793E1/6.0+Thetab/2.0))/sqrt(power(-
cos(ax)*cos(ay)*Rp*sin(0.3141592653589793E1/6.0+Thetap/2.0)-(-
sin(ax)*cos(az)+cos(ax)*sin(ay)*sin(az))*Rp*cos(0.3141592653589793E1/6.0+Thetap/2.0)+x+Rb*si
n(0.3141592653589793E1/6.0+Thetab/2.0),2.0)+power(-
sin(ax)*cos(ay)*Rp*sin(0.3141592653589793E1/6.0+Thetap/2.0)-
(cos(ax)*cos(az)+sin(ax)*sin(ay)*sin(az))*Rp*cos(0.3141592653589793E1/6.0+Thetap/2.0)+y+Rb*c
os(0.3141592653589793E1/6.0+Thetab/2.0),2.0)+power(sin(ay)*Rp*sin(0.3141592653589793E1/6.0
+Thetap/2.0)-cos(ay)*sin(az)*Rp*cos(0.3141592653589793E1/6.0+Thetap/2.0)+z,2.0));
```

j13

```
j13 = (sin(ay)*Rp*sin(0.3141592653589793E1/6.0+Thetap/2.0)-cos(ay)*sin(az)*Rp*
cos(0.3141592653589793E1/6.0+Thetap/2.0)+z)/sqrt(power(-cos(ax)*cos(ay)*Rp*
sin(0.3141592653589793E1/6.0+Thetap/2.0)-(-sin(ax)*cos(az)+cos(ax)*sin(ay)*sin(az))*Rp*
cos(0.3141592653589793E1/6.0+Thetap/2.0)+x+Rb*sin(0.3141592653589793E1/6.0+Thetab/2.0),2.0
)+power(-sin(ax)*cos(ay)*Rp*sin(0.3141592653589793E1/6.0+Thetap/2.0)-
(cos(ax)*cos(az)+sin(ax)*sin(ay)*sin(az))*Rp*cos(0.3141592653589793E1/6.0+Thetap/2.0)+y+Rb*c
```

os(0.3141592653589793E1/6.0+Thetab/2.0),2.0)+power(sin(ay)*Rp*sin(0.3141592653589793E1/6.0
+Thetap/2.0)-cos(ay)*sin(az)*Rp*cos(0.3141592653589793E1/6.0+Thetap/2.0)+z,2.0));

j14

j14 = ((((-cos(ax)*cos(ay)*Rp*sin(0.3141592653589793E1/6.0+Thetap/2.0)-(-
sin(ax)*cos(az)+cos(ax)*sin(ay)*sin(az))*Rp*cos(0.3141592653589793E1/6.0+Thetap/2.0)+x+Rb*si
n(0.3141592653589793E1/6.0+Thetab/2.0))/sqrt(power(-
cos(ax)*cos(ay)*Rp*sin(0.3141592653589793E1/6.0+Thetap/2.0)-(-
sin(ax)*cos(az)+cos(ax)*sin(ay)*sin(az))*Rp*cos(0.3141592653589793E1/6.0+Thetap/2.0)+x+Rb*si
n(0.3141592653589793E1/6.0+Thetab/2.0),2.0)+power(-
sin(ax)*cos(ay)*Rp*sin(0.3141592653589793E1/6.0+Thetap/2.0)-
(cos(ax)*cos(az)+sin(ax)*sin(ay)*sin(az))*Rp*cos(0.3141592653589793E1/6.0+Thetap/2.0)+y+Rb*c
os(0.3141592653589793E1/6.0+Thetab/2.0),2.0)+power(sin(ay)*Rp*sin(0.3141592653589793E1/6.0
+Thetap/2.0)-
cos(ay)*sin(az)*Rp*cos(0.3141592653589793E1/6.0+Thetap/2.0)+z,2.0))))*(sin(ax)*cos(ay)*Rp*sin(
0.3141592653589793E1/6.0+Thetap/2.0)-(-cos(ax)*cos(az)-
sin(ax)*sin(ay)*sin(az))*Rp*cos(0.3141592653589793E1/6.0+ Thetap/2.0)))+(((-
sin(ax)*cos(ay)*Rp*sin(0.3141592653589793E1/6.0+Thetap/2.0)-
(cos(ax)*cos(az)+sin(ax)*sin(ay)*sin(az))*Rp*cos(0.3141592653589793E1/6.0+Thetap/2.0)+y+Rb*c
os(0.3141592653589793E1/6.0+Thetab/2.0))/sqrt(power(-
cos(ax)*cos(ay)*Rp*sin(0.3141592653589793E1/6.0+Thetap/2.0)-(-
sin(ax)*cos(az)+cos(ax)*sin(ay)*sin(az))*Rp*cos(0.3141592653589793E1/6.0+Thetap/2.0)+x+Rb*si
n(0.3141592653589793E1/6.0+Thetab/2.0),2.0)+power(-
sin(ax)*cos(ay)*Rp*sin(0.3141592653589793E1/6.0+Thetap/2.0)-
(cos(ax)*cos(az)+sin(ax)*sin(ay)*sin(az))*Rp*cos(0.3141592653589793E1/6.0+Thetap/2.0)+y+Rb*c
os(0.3141592653589793E1/6.0+Thetab/2.0),2.0)+power(sin(ay)*Rp*sin(0.3141592653589793E1/6.0
+Thetap/2.0)-cos(ay)*sin(az)*Rp*cos(0.3141592653589793E1/6.0+Thetap/2.0)+z,2.0))))* (-
cos(ax)*cos(ay)*Rp*sin(0.3141592653589793E1/6.0+Thetap/2.0)-(-
sin(ax)*cos(az)+cos(ax)*sin(ay)*sin(az))* Rp*cos(0.3141592653589793E1/6.0+Thetap/2.0));

j15

j15 = (((-cos(ax)*cos(ay)*Rp*sin(0.3141592653589793E1/6.0+Thetap/2.0)-(-
sin(ax)*cos(az)+cos(ax)*sin(ay)*sin(az))*Rp*cos(0.3141592653589793E1/6.0+Thetap/2.0)+x+Rb*si
n(0.3141592653589793E1/6.0+Thetab/2.0))/sqrt(power(-
cos(ax)*cos(ay)*Rp*sin(0.3141592653589793E1/6.0+Thetap/2.0)-(-
sin(ax)*cos(az)+cos(ax)*sin(ay)*sin(az))*Rp*cos(0.3141592653589793E1/6.0+Thetap/2.0)+x+Rb*si
n(0.3141592653589793E1/6.0+Thetab/2.0),2.0)+power(-
sin(ax)*cos(ay)*Rp*sin(0.3141592653589793E1/6.0+Thetap/2.0)-
(cos(ax)*cos(az)+sin(ax)*sin(ay)*sin(az))*Rp*cos(0.3141592653589793E1/6.0+Thetap/2.0)+y+Rb*c
os(0.3141592653589793E1/6.0+Thetab/2.0),2.0)+power(sin(ay)*Rp*sin(0.3141592653589793E1/6.0
+Thetap/2.0)-cos(ay)*sin(az)*Rp*cos(0.3141592653589793E1/6.0+Thetap/2.0)+z,2.0))))*
(cos(ax)*sin(ay)*Rp*sin(0.3141592653589793E1/6.0+Thetap/2.0)-cos(ax)*cos(ay)*sin(az)*Rp*
cos(0.3141592653589793E1/6.0+Thetap/2.0))+(((((-
sin(ax)*cos(ay)*Rp*sin(0.3141592653589793E1/6.0+ Thetap/2.0)-
(cos(ax)*cos(az)+sin(ax)*sin(ay)*sin(az))*Rp*cos(0.3141592653589793E1/6.0+Thetap/2.0)+
y+Rb*cos(0.3141592653589793E1/6.0+Thetab/2.0))/sqrt(power(-
cos(ax)*cos(ay)*Rp*sin(0.3141592653589793E1/6.0+Thetap/2.0)-(-
sin(ax)*cos(az)+cos(ax)*sin(ay)*sin(az))*Rp*cos(0.3141592653589793E1/6.0+Thetap/2.0)+x+Rb*si
n(0.3141592653589793E1/6.0+Thetab/2.0),2.0)+power(-
sin(ax)*cos(ay)*Rp*sin(0.3141592653589793E1/6.0+Thetap/2.0)-
(cos(ax)*cos(az)+sin(ax)*sin(ay)*sin(az))*Rp*cos(0.3141592653589793E1/6.0+Thetap/2.0)+y+Rb*c
os(0.3141592653589793E1/6.0+Thetab/2.0),2.0)+power(sin(ay)*Rp*sin(0.3141592653589793E1/6.0
+Thetap/2.0)-cos(ay)*sin(az)*Rp*cos(0.3141592653589793E1/6.0+Thetap/2.0)+z,2.0))))*

```
(sin(ax)*sin(ay)*Rp*sin(0.3141592653589793E1/6.0+Thetap/2.0)-
cos(ay)*sin(az)*sin(ax)*Rp*cos(0.3141592653589793E1/6.0+Thetap/2.0)))+
(((sin(ay)*Rp*sin(0.3141592653589793E1/6.0+Thetap/2.0)-cos(ay)*sin(az)*Rp*
os(0.3141592653589793E1/6.0+Thetap/2.0)+z)/sqrt(power(-
cos(ax)*cos(ay)*Rp*sin(0.3141592653589793E1/6.0+ hetap/2.0)-(-
sin(ax)*cos(az)+cos(ax)*sin(ay)*sin(az))*Rp*cos(0.3141592653589793E1/6.0+Thetap/2.0)+
+Rb*sin(0.3141592653589793E1/6.0+Thetab/2.0),2.0)+power(-
sin(ax)*cos(ay)*Rp*sin(0.3141592653589793E1/6.0+Thetap/2.0)-
(cos(ax)*cos(az)+sin(ax)*sin(ay)*sin(az))*Rp*cos(0.3141592653589793E1/6.0+Thetap/2.0)+y+Rb*c
os(0.3141592653589793E1/6.0+Thetab/2.0),2.0)+power(sin(ay)*Rp*sin(0.3141592653589793E1/6.0
+Thetap/2.0)-
cos(ay)*sin(az)*Rp*cos(0.3141592653589793E1/6.0+Thetap/2.0)+z,2.0)))*(cos(ay)*Rp*sin(0.31415
92653589793E1/6.0+Thetap/2.0)+sin(az)*sin(ay)*Rp*cos(0.3141592653589793E1/6.0+Thetap/2.0)))
);
```

j16

```
j16 = (-((-cos(ax)*cos(ay)*Rp*sin(0.3141592653589793E1/6.0+Thetap/2.0)-(-
sin(ax)*cos(az)+cos(ax)*sin(ay)*sin(az))*Rp*cos(0.3141592653589793E1/6.0+Thetap/2.0)+x+Rb*si
n(0.3141592653589793E1/6.0+Thetab/2.0))/sqrt(power(-
cos(ax)*cos(ay)*Rp*sin(0.3141592653589793E1/6.0+Thetap/2.0)-(-
sin(ax)*cos(az)+cos(ax)*sin(ay)*sin(az))*Rp*cos(0.3141592653589793E1/6.0+Thetap/2.0)+x+Rb*si
n(0.3141592653589793E1/6.0+Thetab/2.0),2.0)+power(-
sin(ax)*cos(ay)*Rp*sin(0.3141592653589793E1/6.0+Thetap/2.0)-
(cos(ax)*cos(az)+sin(ax)*sin(ay)*sin(az))*Rp*cos(0.3141592653589793E1/6.0+Thetap/2.0)+y+Rb*c
os(0.3141592653589793E1/6.0+Thetab/2.0),2.0)+power(sin(ay)*Rp*sin(0.3141592653589793E1/6.0
+Thetap/2.0)-cos(ay)*sin(az)*Rp*cos(0.3141592653589793E1/6.0+Thetap/2.0)+z,2.0))))*
(sin(ax)*sin(az)+cos(ax)*sin(ay)*cos(az))*Rp*cos(0.3141592653589793E1/6.0+Thetap/2.0)+ (- ((-
sin(ax)*cos(ay)*Rp*sin(0.3141592653589793E1/6.0+Thetap/2.0)-
(cos(ax)*cos(az)+sin(ax)*sin(ay)*sin(az))*Rp*cos(0.3141592653589793E1/6.0+Thetap/2.0)+y+Rb*c
os(0.3141592653589793E1/6.0+Thetab/2.0))/sqrt(power(-
cos(ax)*cos(ay)*Rp*sin(0.3141592653589793E1/6.0+Thetap/2.0)-(-
sin(ax)*cos(az)+cos(ax)*sin(ay)*sin(az))*Rp*cos(0.3141592653589793E1/6.0+Thetap/2.0)+x+Rb*si
n(0.3141592653589793E1/6.0+Thetab/2.0),2.0)+power(-
sin(ax)*cos(ay)*Rp*sin(0.3141592653589793E1/6.0+Thetap/2.0)-
(cos(ax)*cos(az)+sin(ax)*sin(ay)*sin(az))*Rp*cos(0.3141592653589793E1/6.0+Thetap/2.0)+y+Rb*c
os(0.3141592653589793E1/6.0+Thetab/2.0),2.0)+power(sin(ay)*Rp*sin(0.3141592653589793E1/6.0
+Thetap/2.0)-cos(ay)*sin(az)*Rp*cos(0.3141592653589793E1/6.0+Thetap/2.0)+z,2.0))))*(-
cos(ax)*sin(az)+sin(ax)*sin(ay)*cos(az))*Rp*cos(0.3141592653589793E1/6.0+Thetap/2.0)+(-
((sin(ay)*Rp*sin(0.3141592653589793E1/6.0+Thetap/2.0)-
cos(ay)*sin(az)*Rp*cos(0.3141592653589793E1/6.0+Thetap/2.0)+z)/sqrt(power(-
cos(ax)*cos(ay)*Rp*sin(0.3141592653589793E1/6.0+Thetap/2.0)-(-
sin(ax)*cos(az)+cos(ax)*sin(ay)*sin(az))*Rp*cos(0.3141592653589793E1/6.0+Thetap/2.0)+x+Rb*si
n(0.3141592653589793E1/6.0+Thetab/2.0),2.0)+power(-
sin(ax)*cos(ay)*Rp*sin(0.3141592653589793E1/6.0+Thetap/2.0)-
(cos(ax)*cos(az)+sin(ax)*sin(ay)*sin(az))*Rp*cos(0.3141592653589793E1/6.0+Thetap/2.0)+y+Rb*c
os(0.3141592653589793E1/6.0+Thetab/2.0),2.0)+power(sin(ay)*Rp*sin(0.3141592653589793E1/6.0
+Thetap/2.0)-
cos(ay)*sin(az)*Rp*cos(0.3141592653589793E1/6.0+Thetap/2.0)+z,2.0)))*cos(ay)*cos(az)*Rp*cos(
0.3141592653589793E1/6.0+Thetap/2.0));
```

MatLab code for getting symbolic presentation of hexapod Jacobian

```
% Define Rb, Rp - radii of the base and platform,
```

```
% Thetab, Thetap - angles between the first two joints on the base and platform
syms Rb Rp Thetab Thetap;

% For later tests use Rb = 7.6; Rp = 3; psib = 30; psip = 0; Thetab = psib/180*pi; Thetap = psip/180*pi;
% psib = 30 and psip = -60 means that the manipulator has hexapod base and triangle platform

Lambdab = zeros(1,6); Lambdap = zeros(1,6);

% Define base angles
baseAngles = 2*pi/3*[ -1 -1  0  0  +1 +1 ];
offset     = [ -1 +1 -1 +1  -1 +1 ];

% Define intermediate angles for further calculations
Lambdab = baseAngles + Thetab/2*offset;
Lambdap = baseAngles + Thetap/2*offset;

base = [ Rb * cos(Lambdab) ; Rb * sin(Lambdab) ; zeros(1,6) ];
plat = [ Rp * cos(Lambdap) ; Rp * sin(Lambdap); zeros(1,6) ];

% Model of a hexapod with 6 joints
model = [base plat];

% b - 4x6 matrix of coordinates of base joints (x y z 1)
b = [ model(:,1:6 ); ones(1,6) ];

% p - 4x6 matrix of coordinates of platform joints
p = [ model(:,7:12); ones(1,6) ];

syms x y z ax ay az;
% 6x1 vector describing the platform position (x,y,z,roll,pitch,yaw)
trpy = [x; y; z; ax; ay; az];

% Rotation matrix
R = SPtrpy2tr(trpy);

% Inverse kinematics
q = R*p - b;

% Find derivative
dR = SPdiffTr(trpy);
dq = dR * p;

% Length of hexapod legs
l = sqrt(q(1,:).^2 + q(2,:).^2 + q(3,:).^2);

% Derivative of dG/dl where G is the function defining leg lengths using Cartesian coordinates
dG = [ q(1:3,:) .* [ 1./l ; 1./l; 1./l]; ones(1,6) ];

% Reshape matrices to get the Jacobian
for i = 1:6,
   G(i,:) =  dG(:,i)' * reshape(dq(:,i), 4,6);
end;

% Get the symbolic presentation of Jacobian matrix element
```

```
  str = ccode(G(1,1));

function R = SPtrpy2tr(x)

  R = transl(x(1:3)) * rpy2tr(x(4:6));

return

function dR = SPdiffTr(x)
% SPdiffTr - returns derivative of a rotation matrix
%
%     dR = SPdiffTr(x)
%
%     x - 6x1 vector containing translation and rotation parameters
%     x = [ Tx Ty Tz roll pitch yaw ]'
%
%     dR - 24x4 matrix, divided into 6 4x4 matrices

cax = cos(x(6)); sax = sin(x(6));
cay = cos(x(5)); say = sin(x(5));
caz = cos(x(4)); saz = sin(x(4));

dR = [
% dR/dTx
0 0 0 1
0 0 0 0
0 0 0 0
0 0 0 0

% dR/dTy
0 0 0 0
0 0 0 1
0 0 0 0
0 0 0 0

% dR/dTz
0 0 0 0
0 0 0 0
0 0 0 1
0 0 0 0

% dR/droll
-cay*saz  -cax*caz-sax*say*saz  caz*sax-cax*say*saz  0
 cay*caz   caz*sax*say-cax*saz  cax*caz*say+sax*saz  0
 0         0         0          0
 0         0         0          0

% dR/dpitch
-caz*say      cay*caz*sax   cax*cay*caz   0
-say*saz      cay*sax*saz   cax*cay*saz   0
-cay         -sax*say      -cax*say       0
 0              0             0           0

% dR/dyaw
0  cax*caz*say+sax*saz     -caz*sax*say+cax*saz  0
```

```
0 -caz*sax+cax*say*saz     -cax*caz-sax*say*saz  0
0  cax*cay                 -cay*sax              0
0 0                           0                  0
];
return
```

MatLab code for minimization of 6x6 matrix infinity norm using BFGS algorithm

```
function hexapod_bfgs_yv(action)
%============================================================================
===
%HEXAPOD_BFGS_YV - Minimization of hexapod Jacobian inverse using BFGS algorithm
%
%  This MatLab code implements the minimization of hexapod Jacobian inverse
%  using BFGS algorithm
%
%  f(x)= norm(inv(J),inf)
%
%  $Revision: 0.0.0.1 $  $Date: 2006/01/08 $
%  Author: Yauheni Veryha
%============================================================================
===

% Ensure that no arguments are provided during initialization
if nargin<1,
  action='initialize';
end;

% Initialization of GUI
if strcmp(action,'initialize'),
  figNumber=figure( ...
    'Name','Minimization of hexapod Jacobian inverse using BFGS algorithm', ...
    'NumberTitle','off', ...
    'Visible','off');
  axes( ...
    'Units','normalized', ...
    'Visible','off', ...
    'Position',[0.05 0.40 0.70 0.75]);

  % Set up the Comment Window
  top=0.95;
  left=0.05;
  right=0.75;
  bottom=0.05;
  labelHt=0.05;
  spacing=0.005;
  promptStr=str2mat(' ',...
    ' Minimization of hexapod Jacobian inverse', ...
    ' ', ...
    ' f(x)= norm(inv(J),inf)', ...
    ' ', ...
    ' using BFGS algorithm. Press button at the right to start ...');
  % First, the MiniCommand Window frame
  frmBorder=0.02;
```

```
frmPos=[left-frmBorder bottom-frmBorder ...
    (right-left)+2*frmBorder (top-bottom)+2*frmBorder];
uicontrol( ...
  'Style','frame', ...
  'Units','normalized', ...
  'Position',frmPos, ...
  'BackgroundColor',[0.5 0.5 0.5]);
% Then the text label
labelPos=[left top-labelHt (right-left) labelHt];
uicontrol( ...
  'Style','text', ...
  'Units','normalized', ...
  'Position',labelPos, ...
  'BackgroundColor',[0.5 0.5 0.5], ...
  'ForegroundColor',[1 1 1], ...
  'String','Comment Window');
% Then the editable text field
txtPos=[left bottom (right-left) top-bottom-labelHt-spacing];
txtHndl=uicontrol( ...
  'Style','edit', ...
  'HorizontalAlignment','left', ...
  'Units','normalized', ...
  'Max',10, ...
  'BackgroundColor',[1 1 1], ...
  'Position',txtPos, ...
  'Callback','hexapod_bfgs_yv(''eval'')', ...
  'String',promptStr);
% Save this handle for future use
set(gcf,'UserData',txtHndl);

%====================================
% Information for all buttons
labelColor=[0.8 0.8 0.8];
top=0.95;
left=0.80;
btnWid=0.15;
btnHt=0.08;
% Spacing between the button and the next command's label
spacing=0.02;

%====================================
% The CONSOLE frame
frmBorder=0.02;
yPos=0.05-frmBorder;
frmPos=[left-frmBorder yPos btnWid+2*frmBorder 0.9+2*frmBorder];
uicontrol( ...
  'Style','frame', ...
  'Units','normalized', ...
  'Position',frmPos, ...
  'BackgroundColor',[0.5 0.5 0.5]);

%====================================
% "Start optimization using BFGS algorithm" button
btnNumber=1;
yPos=top-(btnNumber-1)*(btnHt+spacing);
```

```
  labelStr='Start';
  callbackStr='hexapod_bfgs_yv("startbutton")';

  % Generic popup button information
  btnPos=[left yPos-btnHt btnWid btnHt];
  uicontrol( ...
    'Style','pushbutton', ...
    'Units','normalized', ...
    'Position',btnPos, ...
    'String',labelStr, ...
    'Callback',callbackStr, ...
    'UserData',btnNumber);

  %====================================
  % The CLOSE button
  uicontrol( ...
    'Style','push', ...
    'Units','normalized', ...
    'Position',[left bottom btnWid btnHt], ...
    'String','Close', ...
    'Callback','close(gcf)');

  % Now uncover the figure
  set(figNumber,'Visible','on');

% This check is for callback function when "Start" button was pressed
elseif strcmp(action,'startbutton'),
  method=get(gco,'UserData');
  txtHndl=get(gcf,'UserData');

  % Initial end-effector coordinates
  x=[-0.02 -0.01 0.26 0.1 0.05 -0.07];

  j11='((-cos(x(4))*cos(x(5))*0.03*0-(-
sin(x(4))*cos(x(6))+cos(x(4))*sin(x(5))*sin(x(6)))*0.03*1+x(1)+0.076*sin(0.5236+0.5236/2.0))/sqrt(
power(-cos(x(4))*cos(x(5))*0.03*0-(-
sin(x(4))*cos(x(6))+cos(x(4))*sin(x(5))*sin(x(6)))*0.03*1+x(1)+0.076*sin(0.5236+0.5236/2.0),2.0)+
power(-sin(x(4))*cos(x(5))*0.03*0-
(cos(x(4))*cos(x(6))+sin(x(4))*sin(x(5))*sin(x(6)))*0.03*1+x(2)+0.076*cos(0.5236+0.5236/2.0),2.0)
+power(sin(x(5))*0.03*0-cos(x(5))*sin(x(6))*0.03*1+ x(3),2.0)))';

  j12 = ' ((-sin(x(4))*cos(x(5))*0.03*0-
(cos(x(4))*cos(x(6))+sin(x(4))*sin(x(5))*sin(x(6)))*0.03*1+x(2)+0.076*cos(0.5236+0.5236/2.0))/sqrt
(power(-cos(x(4))*cos(x(5))*0.03*0-(-
sin(x(4))*cos(x(6))+cos(x(4))*sin(x(5))*sin(x(6)))*0.03*1+x(1)+0.076*sin(0.5236+0.5236/2.0),2.0)+
power(-sin(x(4))*cos(x(5))*0.03*0-
(cos(x(4))*cos(x(6))+sin(x(4))*sin(x(5))*sin(x(6)))*0.03*1+x(2)+0.076*cos(0.5236+0.5236/2.0),2.0)
+power(sin(x(5))*0.03*0-cos(x(5))*sin(x(6))*0.03*1+x(3),2.0)))';

  j13 = ' ((sin(x(5))*0.03*0-cos(x(5))*sin(x(6))*0.03*1+x(3))/sqrt(power(-
cos(x(4))*cos(x(5))*0.03*0-(-
sin(x(4))*cos(x(6))+cos(x(4))*sin(x(5))*sin(x(6)))*0.03*1+x(1)+0.076*sin(0.5236+0.5236/2.0),2.0)+
power(-sin(x(4))*cos(x(5))*0.03*0-
(cos(x(4))*cos(x(6))+sin(x(4))*sin(x(5))*sin(x(6)))*0.03*1+x(2)+0.076*cos(0.5236+0.5236/2.0),2.0)
+power(sin(x(5))*0.03*0-cos(x(5))*sin(x(6))*0.03*1+x(3),2.0)))';
```

% We do not show all elements of hexapod Jacobian because of limited space

```
  % Define function for minimization, i.e., infinity norm of 6 x 6 matrix
  strBegin = 'norm(inv( [';
  strEnd = '] ),inf)';
  strMiddle = ';';

  f = strcat(strBegin, j11, j12, j13, j14, j15, j16, strMiddle, j21, j22, j23, j24, j25, j26, strMiddle, j31,
j32, j33, j34, j35, j36, strMiddle, j41, j42, j43, j44, j45, j46, strMiddle, j51, j52, j53, j54, j55, j56,
strMiddle, j61, j62, j63, j64, j65, j66, strEnd);

  OPTIONS=optimset('LargeScale','off','OutputFcn',@hexapod_bfgs_yvOutFcn);

  % Turn off all messages
  OPTIONS = optimset(OPTIONS,'display','off');

  % Broyden-Fletcher-Goldfarb-Shanno method implemented in MatLab
  if method==1,
     str= ...
       [' Broyden-Fletcher-Goldfarb-Shanno          '
        ' (Unconstrained quasi-Newton minimization)'];
     set(txtHndl,'String',str);

     % No gradient because the function is too complex to find its derivatives
     OPTIONS = optimset(OPTIONS,'gradobj','off','MaxFunEvals',1000, 'MaxIter',200, ...
        'InitialHessType','scaled-identity');

     str2=' [x,fval,exitflag,output] = fminunc(f,x,OPTIONS);';
     [x,fval,exitflag,output] = fminunc(f,x,OPTIONS);
  end;

  % Prepare final optimization results
  funEvals=sprintf(' Number of iterations: %g.  Number of function evaluations: %g.',
output.iterations, output.funcCount);
  str=get(txtHndl,'String');
  str=str2mat(str, str2,' ', ' Results of hexapod Jacobian minimization can be found in
Hexapod_BFGS_Results.xls file!', ...
      ' ', ' f(x)= norm(inv(J),inf)', funEvals);
  set(txtHndl,'String',str);

end;    % if strcmp(action, ...

%-----------------------------------------------------------------
% Callback after one iteration step
function stop = hexapod_bfgs_yvOutFcn(x,optimvalues,state,userdata,varargin)

% Do not stop minimization at this point
stop = false;

% Evaluate current values and output function values to MS Excel
if strcmp(state,'iter')

 % Actual function value, i.e., infinity norm of 6 x 6 matrix
```

```
Fi = norm(inv([((-cos(x(4))*cos(x(5))*0.03*0-(-
sin(x(4))*cos(x(6))+cos(x(4))*sin(x(5))*sin(x(6)))*0.03*1+x(1)+0.076*sin(0.5236+0.5236/2.0))/sqrt(
power(-cos(x(4))*cos(x(5))*0.03*0-(-
sin(x(4))*cos(x(6))+cos(x(4))*sin(x(5))*sin(x(6)))*0.03*1+x(1)+0.076*sin(0.5236+0.5236/2.0),2.0)+
power(-sin(x(4))*cos(x(5))*0.03*0-
(cos(x(4))*cos(x(6))+sin(x(4))*sin(x(5))*sin(x(6)))*0.03*1+x(2)+0.076*cos(0.5236+0.5236/2.0),2.0)
+power(sin(x(5))*0.03*0-cos(x(5))*sin(x(6))*0.03*1+ x(3),2.0))) ...
    ((-sin(x(4))*cos(x(5))*0.03*0-
(cos(x(4))*cos(x(6))+sin(x(4))*sin(x(5))*sin(x(6)))*0.03*1+x(2)+0.076*cos(0.5236+0.5236/2.0))/sqrt
(power(-cos(x(4))*cos(x(5))*0.03*0-(-
sin(x(4))*cos(x(6))+cos(x(4))*sin(x(5))*sin(x(6)))*0.03*1+x(1)+0.076*sin(0.5236+0.5236/2.0),2.0)+
power(-sin(x(4))*cos(x(5))*0.03*0-
(cos(x(4))*cos(x(6))+sin(x(4))*sin(x(5))*sin(x(6)))*0.03*1+x(2)+0.076*cos(0.5236+0.5236/2.0),2.0)
+power(sin(x(5))*0.03*0-cos(x(5))*sin(x(6))*0.03*1+x(3),2.0))) ...
    ((sin(x(5))*0.03*0-cos(x(5))*sin(x(6))*0.03*1+x(3))/sqrt(power(-cos(x(4))*cos(x(5))*0.03*0-(-
sin(x(4))*cos(x(6))+cos(x(4))*sin(x(5))*sin(x(6)))*0.03*1+x(1)+0.076*sin(0.5236+0.5236/2.0),2.0)+
power(-sin(x(4))*cos(x(5))*0.03*0-
(cos(x(4))*cos(x(6))+sin(x(4))*sin(x(5))*sin(x(6)))*0.03*1+x(2)+0.076*cos(0.5236+0.5236/2.0),2.0)
+power(sin(x(5))*0.03*0-cos(x(5))*sin(x(6))*0.03*1+x(3),2.0))) ...
    % We do not show all elements of hexapod Jacobian because of limited space
    ] ),inf);

  % Output the header only in the beginning
  if optimvalues.iteration == 0,
    mHeader = {'i', 'x(1)', 'x(2)', 'x(3)', 'x(4)','x(5)', 'x(6)', 'F'};
    xlswrite('Hexapod_BFGS_Results', mHeader, 'Sheet1','A1')
  end;

  % optimvalues.iteration > 0 is used to avoid double entries in the first function evaluation
  if optimvalues.iteration > 0,
    % Output matrix with optimization results on the current iteration
    mOutput = {optimvalues.iteration, x(1), x(2), x(3), x(4), x(5), x(6), Fi};
    xlswrite('Hexapod_BFGS_Results', mOutput, 'Sheet1', ['A' int2str(optimvalues.iteration + 1)])
  end;
end
```